Y0-BQE-572

1,001 Chemicals in Everyday Products

1,001 Chemicals in Everyday Products

Grace Ross Lewis

VAN NOSTRAND REINHOLD
New York

Library of Congress Catalog Card Number 93-29791
ISBN 0-442-01458-9

I(T)P Van Nostrand Reinhold is an International Thomson Publishing company.
ITP logo is a trademark under license.

Printed in the United States of America

Van Nostrand Reinhold
115 Fifth Avenue
New York, NY 10003

International Thomson Publishing GmbH
Königswinterer Str. 418
53227 Bonn
Germany

International Thomson Publishing
Berkshire House,168-173
High Holborn, London WC1V 7AA
England

International Thomson Publishing Asia
221 Henderson Bldg. #05-10
Singapore 0315

Thomas Nelson Australia
102 Dodds Street
South Melbourne 3205
Victoria, Australia

International Thomson Publishing Japan
Kyowa Building, 3F
2-2-1 Hirakawacho
Chiyoda-Ku, Tokyo 102
Japan

Nelson Canada
1120 Birchmount Road
Scarborough, Ontario
M1K 5G4, Canada

16 15 14 13 12 11 10 9 8 7 6 5 4

Library of Congress Cataloging-in-Publication Data

Lewis, Grace Ross.
1,001 chemicals in everyday products / Grace Ross Lewis.
p. cm.
Includes index.
ISBN 0-442-01458-9
1. Chemicals. I. Title. II. One thousand and one chemicals in everyday products. III. One thousand one chemicals in everyday products.
TP200.L49 1994
363.17'91—dc20 93-29791
CIP

To Dick
for 1,001 reasons

CONTENTS

PREFACE

"Knowledge is of two kinds. We know a subject ourselves or we know where we can find information upon it." Samuel Johnson

We live in a chemical soup of pollution. Nearly everyone uses cleaning products and deodorants and consumes food with artificial coloring and preservatives. We are surrounded by the outgassing of formaldehyde from furniture and flooring, chemicals released by freshly cleaned clothing, and even backyard grilling. We are all exposed.

Since everything is composed of chemicals we have to realize that there are two sides to this story. Many, if not most chemicals, when used in appropriate amounts for the correct use, are beneficial to us. The EPA states that trace levels of pesticide residues in food pose no health hazard. Others embrace the Delaney Clause, named for Representative Delaney of New York, who introduced it into legislature. This Clause requires the complete prohibition of residues of any additive that has been shown to cause cancer in humans or lab animals.

Almost everyone has had some type of allergic reaction to a food, toiletry, or cleaning product. We have all experienced at least a minor skin rash after eating or using something new. For victims of Multiple Chemical Sensitivities (MCS), just living, eating and working is a challenge.

Problems do develop with excessive exposure and inappropriate uses. That is why labeling is so important. We can protect ourselves from unnecessary chemical exposure only if we have the facts.

A great deal of health and safety information on chemicals is available. This information (found in books, microfiche, CD-ROMs, and databases) is used primarily by industrial hygienists, toxicologists, occupational health physicians, nurses, safety officers, and environmental scientists. This book contains information extracted from these professional references. The facts about these chemicals were drawn from studies where pure, concentrated samples where tested. In many cases, substances are diluted by the time consumers come in contact with them.

In this book I have tried to briefly mention the benefits and hazards of many commonly-used chemicals, as well as interesting uses, to allow you to make informed decisions about the products you use and to add to your knowledge of everyday chemicals.

California has a chemical and product labeling law that requires manufacturers to disclose the ingredients in a product which may cause cancer or birth defects. As a result of this legislation, manufacturers reformulated many products that contained carcinogenic (cancer-causing) and toxic chemicals. While the public would benefit from this type of law in every state, industry has prevented its passage in many areas.

Many demand that the Government not allow products to come to market if they can be harmful. At the same time, few of us are willing to pay additional taxes to pay for more government scientists to conduct testing. It is a dilemma.

Ultimately, we are all responsible for our own well being.

HOW TO USE THIS BOOK

1001 Chemicals in Everyday Products provides information about those mysterious-sounding chemicals listed on the labels of household products. It contains information on over 1,000 chemicals. The following example illustrates how an ingredient in hand soap is presented in this book.

TRICLOSAN

Products: In deodorants, hand soaps, vaginal deodorants, cosmetics, pharmaceuticals, and cleansers.

Use: A disinfectant/antibacterial common in consumer toiletries.

Precautions: Moderately toxic by swallowing. Mildly toxic by skin contact. A human mutagen (changes inherited characteristics). A skin irritant and possible allergen.

Synonyms: CAS: 3380-34-5 ✲ IRGASAN ✲ 5-CHLORO-2-(2,4-DICHLOROPHENOXYPHENOL)

You will find that every chemical names a few *Products* that contain the chemical. For instance, under "Triclosan" you will find the products "deodorants, handsoaps, vaginal deodorants, cosmetics, pharmaceuticals, and cleansers."

The Triclosan listing also has a heading for *Uses* that includes "disinfectant/antibacterial common in consumer toiletries." If you were trying to avoid Triclosan, you would be alerted to check consumer toiletries for its presence.

Another information heading is *Health Effects*. The facts used as the basis for *Health Effects* come from laboratory findings. Of course, this information is not meant to be used as medical advice. Anyone who experiences illness should seek medical advice. The words, toxic and poisonous have various meanings. The results of contact with toxic or poisonous chemicals could produce anything from mild discomfort to death. This depends on the amount and length of contact with the chemicals.

Two important things to consider are how much of a product is used and how often it is used. For instance, a teaspoon of vinegar is safe to use on a salad, but if a person drank several cupfuls daily it could cause serious harm. So we can say 'the dose makes the poison.' In everyday products the amounts used are at safe levels for the average person. The consumer would usually not apply or eat enough of the chemical to be hazardous. However, when misused, used by children in inappropriate amounts, or by people who have particular allergies or sensitivities, there is reason for concern.

The final category is *Synonyms*. The synonyms list includes other names by which the chemical is known. If you are allergic to Triclosan, you would also want to avoid products that contain Irgasan. The synonym list tells us that they are one and the same. The CAS (Chemical Abstracts Service) number is another synonym for the chemical increasingly found on labels.

The illustrations in this book are symbols relating to the *Uses* or *Products* of each chemical. They are not a code and do not necessarily indicate chemical categories.

The index at the back of the book provides an easy way to find the page number for any information contained in the book. You will find listings of all the chemicals, products, uses, health effects, and synonyms. The terms in bold type refer to the more familiar words while the smaller-sized type refers to the chemical names and synonyms.

You will also find a list of the Poison Control Centers for each state. This reference was included especially for parents, teachers, students, and caregivers. These centers serve as a resource for information on toxic substances and can provide round-the-clock assistance in emergency situations.

Chemical research is an ongoing activity, with new information available every day. As a result, this book does not indicate everything known about every chemical. Not all chemicals, products, and uses are itemized. More information on each chemical is available in books like those listed in this book's bibliography, via on-line computer databases like InfoTrac and Medline, and from Poison Control Centers, or your public or university library.

Grace Ross Lewis

ACKNOWLEDGEMENTS

I want to thank Bob Esposito, Alex Padro, Michael Sherry, Peter Rocheleau and Caroline McCarra of Van Nostrand Reinhold for their assistance in the preparation of this book.

I am grateful for the suggestions and partial review of manuscript by Julie R. Lewis, Esq., Richard J. Lewis, Jr., Velma W. Ross, Brenda Bien Young and Doris V. Sweet.

Books of Related Interest from Van Nostrand Reinhold

Chemical Exposures: Low Levels and High Stakes

by Nicholas A. Ashford, Ph.D. and Claudia S. Miller, M. D.

The Dose Makes the Poison, Second Edition

by M. Alice Ottoboni, Ph.D.

Food Additives Handbook

by Richard J. Lewis, Sr.

Hawley's Condensed Chemical Dictionary, Twelfth Edition

Richard J. Lewis, Sr.

In Our Backyard: A Guide to Pollution and It s Effects

by Travis P. Wagner

1,001 Chemicals in Everyday Products

A

"About 12 million Americans have some type of food allergy."
American College of Allergy and Immunology

ABSINTHIUM

Products: In liqueurs or vermouth.

Use: As a flavoring.

Health Effects: Moderately toxic by swallowing large amounts. An allergen. May cause contact dermatitis (skin rash and irritation).

Synonyms: CAS: 8022-37-5 ✲ ARTEMISIA OIL ✲ WORMWOOD

ACENAPTHENE

Products: In garden chemicals and plastics.

Use: As insecticide or fungicide.

Health Effects: Derived from coal tar. A known carcinogen (causes cancer) and allergen.

Synonyms: ✲ ETHYLENENAPTHALENE ✲ 1,8-DIHYDROACENAPTHALENE

ACESULFAME POTASSIUM

Products: In beverage mixes, chewing gum, instant coffee and tea, dry dairy products, dry puddings, desserts, and tabletop sweetener.

Use: As a nonnutritive sweetener (approximately 24% of the sweetener market).

Health Effects: FDA approves use at moderate levels to accomplish the desired results.

Synonyms: CAS: 55589-62-3 ✲ ACESULFAME K ✲ SUNETTE

ACETAL

Products: In cosmetics, jasmine perfumes, and fruit flavors.

Use: As a solvent and odorant.

Health Effects: Highly flammable. Moderately toxic by swallowing large amounts. Narcotic in high concentrations when breathed.

Synonyms: CAS: 105-57-7 ❊ DIETHYLACETAL ❊ 1,2- DIETHOXYETHANE ❊ ETHYLIDENEDIETHYL ETHER

ACETALDEHYDE

Products: In perfume, food, cosmetics, adhesives, and glues.

Use: As a solvent and flavoring.

Health Effects: FDA approves use at moderate levels to accomplish the intended results. It is considered toxic and narcotic in excessive amounts. A skin and eye irritant. Occurs naturally in some fruits and vegetables.

Synonyms: CAS: 75-07-0 ❊ ACETIC ALDEHYDE ❊ ALDEHYDE ❊ ETHANAL ❊ ETHYL ALDEHYDE

ACETATE FIBER, SAPONIFIED

Products: In typewriter ribbons, belts, tapes, and carpet backing.

Use: For strength and durability.

Health Effects: Harmless when used for intended purposes.

Synonyms: ❊ REGENERATED CELLULOSE FIBERS

ACETIC ACID

Products: In baked goods, catsup, cheese, chewing gum, condiments, dairy products, fats, gravies, mayonnaise, meat products, oil, pickles, poultry, relishes, salad dressings, and sauces. In skin bleaching cosmetics and hair coloring products.

Use: As a flavor enhancer, pickling agent, and solvent.

Health Effects: The pure substance in gross amounts is moderately toxic when swallowed and breathed. Diluted material is approved for food use. A strong irritant to skin and tissue. A severe eye irritant. FDA states it is GRAS (generally regarded as safe) within limitations.

Synonyms: CAS: 64-19-7 ❊ ETHANOIC ACID ❊ VINEGAR ACID ❊ METHANECARBOXYLIC ACID ❊ ETHYLIC ACID

ACETIC ACID VINYL ESTER POLYMERS

Products: In chewing gum, confectionery, food supplements, and packaging materials.

Use: As a resin in adhesives, paints, and paper coatings. A color fixative, a masticatory substance, chewing gum, and nonwoven binders.

Health Effects: Harmless when used for appropriate purposes and in moderate amounts.

Synonyms: CAS: 9003-20-7 ✲ ACETIC ACID ETHENYL ✲ ESTER HOMOPOLYMER ✲ BAKELITE AYAA ✲ BOOKSAVER ✲ BORDEN 2123 ✲ DANFIRM ✲ DARATAK ✲ ELMER'S GLUE-ALL ✲ POLYVINYL ACETATE ✲ VINYL ACETATE POLYMER ✲ VINYL ACETATE RESIN

ACETONE

Products: In nail polish, nail polish remover, markers, airplane glue solvent, fabric cement, cleaning fluids, paint, varnish, and lacquer.

Use: As a solvent, also to clean and dry.

Health Effects: Flammable, dangerous fire risk. Moderately toxic by swallowing and breathing. A skin and severe eye irritant. Narcotic effect in high concentrations. Can cause coma, kidney damage, and heart effects.

Synonyms: CAS: 67-64-1 ✲ DIMETHYLKETONE ✲ 2-PROPANONE ✲ KETONE PROPANE ✲ METHYL KETONE ✲ PROPANONE ✲ PYROACETIC ACID ✲ PYROACETIC ETHER

ACETONE CYANOHYDRIN

Products: In insecticides.

Use: As a pesticide.

Health Effects: Toxic. On EPA extremely hazardous substance list.

Synonyms: CAS: 75-86-5 ✲ ALPHA-HYDROXYISOBUTRONITRILE ✲ 2-METHYLLACTONITRILE

N-ACETYL ETHANOLAMINE

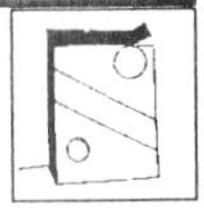

Products: In paper products, glues, cork, inks, and textiles.

Use: As a humidifier and conditioner.

Health Effects: Mildly toxic by swallowing. A skin and eye irritant.

Synonyms: CAS: 142-26-7- ✲ HYDROXYETHYLACETAMIDE. ✲ ACETYL COLAMINE

ACETYL TRIETHYL CITRATE

Products: In packaging materials

Use: As a plasticizer; to keep wrapping materials soft and pliable.

Health Effects: FDA approves use at moderate levels to accomplish the intended results. Moderately toxic by swallowing.

Synonyms: CAS: 77-89-4 ✲ CITRIC ACID ✲ ACETYL TRIETHYL ESTER ✲ TRIETHYL ACETYLCITRATE

ACETYL VALERYL

Products: In beverages, ice cream desserts, candy, bakery goods, gum,

Use: As a flavoring.

cheese, butter, fruit, berries, nuts, and rum.

Health Effects: Harmless when used for intended purpose.

Synonyms: ✲ HEPTADIONE-2,3

ACID HYDROLYZED PROTEINS

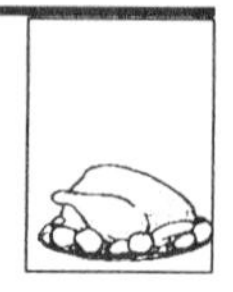

Products: In bologna, salami, sauces, and stuffing.

Use: As a flavoring agent.

Health Effects: USDA approves use in appropriate amounts for designated use.

Synonyms: ✲ HYDROLYZED MILK PROTEIN ✲ HYDROLYZED PLANT PROTEIN ✲ HYDROLYZED VEGETABLE PROTEIN

ACRYLAMIDE

Products: In adhesives, textiles, and permanent-press fabrics. In soil conditioners.

Use: As a coating and conditioner.

Health Effects: Suspected human carcinogen (may cause cancer). Toxic by skin absorption. Irritant to skin, nose, and throat.

Synonyms: CAS: 79-06-1 ✲ ACRYLIC AMIDE ✲ ETHYLENECARBOXAMIDE ✲ PROPENAMIDE

ACTIVATED CARBON

Products: In production and processing of fats, grape juice, cocktail sherry, and wine.

Use: As a decolorizing agent, odor-removing agent, purfication agent, and taste remover.

Health Effects: GRAS (generally regarded as safe) when used appropriately.

Synonyms: CAS: 64365-11-3 ✲ CHARCOAL, ACTIVATED

ADIPIC ACID

Products: In baked goods, baking powder, beverages (nonalcoholic), condiments, dairy product analogs, desserts (frozen dairy), drinks (powdered), fats, gelatins, gravies, margarine, meat products, oil, oil (edible), oleomargarine, puddings, relishes, snack foods, and vegetables (canned).

Use: As a flavoring agent, leavening agent, neutralizing agent, and additive.

Health Effects: In large amounts the pure acid is moderately toxic. A severe eye irritant. Harmless when used for intended purpose in limited amounts.

Synonyms: CAS: 124-04-9 ❊ ACIFLOCTIN ❊ ACINETTEN ❊ ADILACTETTEN ❊ ADIPINIC ACID ❊ 1,4-BUTANEDICARBOXYLIC ACID ❊ 1,6-HEXANEDIOIC ACID ❊ KYSELINA ADIPOVA (CZECH) ❊ MOLTEN ADIPIC ACID

AFLATOXIN

Products: Found in corn or peanuts.

Use: A natural contaminant. Prevention of mold growth is the best protection.

Health Effects: A confirmed human carcinogen (causes cancer) and animal tumorigen (causes tumor growth). A human poison by swallowing. It cannot be totally eliminated. FDA sets allowable limitations.

Synonyms: CAS: 1402-68-2

AGAR

Products: In beverages, baked goods and mixes, candy, confections, frostings, glazes, jellied meats, dental impression material, and laxatives.

Use: As a preservative and gelling agent.

Health Effects: A possible allergen. The concentrated chemical is mildly toxic by swallowing gross amounts. FDA states GRAS (generally regarded as safe) when used within stated limitations in food. Derived from red seaweed.

Synonyms: CAS: 9002-18-0 ❊ AGAR-AGAR ❊ AGAR AGAR FLAKE ❊ AGAR-AGAR GUM ❊ BENGAL GELATIN ❊ BENGAL ISINGLASS ❊ CEYLON ISINGLASS ❊ CHINESE ISINGLASS ❊ DIGENEA SIMPLEX MUCILAGE ❊ GELOSE ❊ JAPAN AGAR ❊ JAPAN ISINGLASS ❊ LAYOR CARANG

ALDOL

Products: In perfume, dyes, and drugs (as a sedative).

Use: As a solvent and various other uses.

Health Effects: A skin irritant in concentrated form.

Synonyms: ❊ ACETADOL ❊ BETA-HYDROXYBUTYRALDEHYDE

ALDRIN

Products: In Insecticides.

Use: In agricultural products such as fertilizers, herbicides, and fungicides.

Health Effects: A poison by swallowing and by skin contact. It has harmful effects on the human body by swallowing. When swallowed it causes confusion, tremors, nausea, or vomiting. Continued exposure causes liver damage. A human mutagen (changes inherited characteristics). Possibly causes cancer in humans.

Synonyms: CAS: 309-00-2 ✼ ALDREX ✼ ALDREX 30 ✼ ALDRIN ✼ ALDRINE (FRENCH) ✼ ALDRITE ✼ ALDROSOL ✼ ALTOX ✼ DRINOX ✼ HHDN ✼ OCTALENE ✼ SEEDRIN

ALGINIC ACID

Products: In soup mixes, soups, antacid ingredient, ice cream, toothpaste, cosmetics, and waterproofing agent for concrete.

Use: As a thickener, emulsifier, or stabilizer.

Health Effects: FDA states it is GRAS (generally regarded as safe).

Synonyms: CAS: 9005-32-7 ✼ KELACID ✼ LANDALGINE ✼ NORGINE ✼ PLOYMANNURONIC ACID ✼ SAZZIO

ALKALOID

Products: In morphine, nicotine, quinine, codeine, caffeine, cocaine, and strychnine.

Use: A stimulant. Caffeine is added to beverages and is found naturally in coffee and tea. Nicotine is naturally occurring in tobacco products and added to pesticides for its toxic qualities.

Health Effects: Poisonous. An allergen.

Synonyms: ✼ ALKALOID SALTS

ALKYL POLYGLYCOSIDES

Products: In detergents, soaps, and shampoos.

Use: As a surfactant (soil loosener) or cleansing agent.

Health Effects: Reported to be biodegradable and less irritating to skin and eyes than similar chemicals.

Synonyms: ✼ GLUCOPON

ALLSPICE

Products: In cakes, fruit pies, mincemeat, plum puddings, sauces, and soups.

Use: As flavoring agent.

Health Effects: A sensitizer which can cause dermatitis after repeated contact. Considered moderately toxic in large amounts.

Synonyms: ⁂ PIMENTA OIL ⁂ PIMENTA BERRIES OIL ⁂ PIMENTO OIL

ALLYL ALCOHOL

Products: A herbicide (kills plants and vegetation).

Use: In garden and agricultural products.

Health Effects: A poison by breathing, swallowing, and skin contact. A skin and severe eye irritant.

Synonyms: CAS: 107-18-6 ⁂ ALLYLIC ALCOHOL ⁂ 3-HYDROXYPROPENE ⁂ ORVINYLCARBINOL ⁂ PROPENOL ⁂ PROPENYL ALCOHOL ⁂ WEED DRENCH

ALLYL BROMIDE

Products: In perfumes.

Use: As a stabilizer.

Health Effects: Strong irritant to skin and eyes in concentrated form.

Synonyms: CAS: 106-95-6 ⁂ BROMALLYLENE ⁂ 3-BROMOPROPENE ⁂ 3-BROMOPROPYLENE

ALLYL CAPROATE

Products: In candy, dessert gels, and puddings.

Use: As a flavoring agent.

Health Effects: FDA approves use at moderate levels to accomplish the intended results. It is a poison if gross amounts of the concentrated chemical are swallowed. An irritant to skin.

Synonyms: CAS: 123-68-2 ⁂ ALLYL HEXANOATE ⁂ 2-PROPENYL-N-HEXANOATE

ALLYL CINNAMATE

Products: In baked goods and candy.

Use: As a flavoring agent.

Health Effects: FDA approves use at moderate levels to accomplish the desired results. Poison by swallowing gross amounts of the pure substance. A skin irritant.

Synonyms: CAS: 1866-31-5 ⁂ ALLYL-3-PHENYLACRYLATE ⁂ PROPENYL CINNAMATE ⁂ VINYL CARBINYL CINNAMATE

ALLYL ISOTHIOCYANATE

Products: In baked goods, condiments, horseradish flavor (imitation), meat,

Use: As a flavoring agent

mustard oil (artificial), pickles, salad dressings, and sauces.

Health Effects: FDA approves use at moderate levels to accomplish the intended results. The pure chemical is poison if excessive amounts are swallowed. Suspected carcinogen (might cause cancer). An eye irritant. An allergen. May cause contact dermatitis.

Synonyms: CAS: 57-06-7 ✲ AITC ✲ ALLYL ISORHODANIDE ✲ ALLYL ISOSULFOCYANATE ✲ ALLYL MUSTARD OIL ✲ ALLYL SEVENOLUM ✲ ALLYL THIOCARBONIMIDE ✲ ARTIFICIAL MUSTARD OIL ✲ CARBOSPOL ✲ 3-ISOTHIOCYANATO-1-PROPENE ✲ MUSTARD OIL ✲ OLEUM SINAPIS VOLATILE ✲ 2-PROPENYL ISOTHIOCYANATE ✲ REDSKIN ✲ VOLATILE OIL of MUSTARD

ALLYL OCTANOATE

Products: In beverages, candy, dessert gels, and puddings.

Use: As a flavoring agent.

Health Effects: FDA approves use at moderate levels to accomplish the intended results. Moderately toxic by swallowing large amounts of pure substance. A skin irritant.

Synonyms: CAS: 4230-97-1 ✲ ALLYL CAPRYLATE ✲ OCTANOIC ACID ALLYL ESTER ✲ OCTANOIC ACID-2-PROPENYL ESTER

ALLYL SULFIDE

Products: Occurs naturally in garlic and horseradish. In fruits, beverages, ice cream, condiments, and meats.

Use: As a flavoring.

Health Effects: An irritant to skin, eyes, nose, and throat

Synonyms: CAS: 592-88-1 ✲ ALLYL MONOSULFIDE ✲ DIALLYL SULFIDE ✲ DIALLYL THIOETHER ✲ OIL GARLIC ✲ THIOALLYL ETHER

ALLYL TRICHLOROSILANE

Products: In boat and car repair products. Produces a fiber glass finish.

Use: Component of fiber glass material.

Health Effects: A fire hazard. Possible skin and eye irritant.

Synonyms: ✲ TRICHLOROALLYLSILANE

ALOE VERA

Products: Skin lotions, creams, ointments, soaps, makeup, moisturizers and (OTC) over the counter burn medications.

Use: As a cosmetic and pharmaceutical ingredient derived from the leaves of a tropical succulent houseplant.

Health Effects: Could cause allergic reaction in susceptible individuals. Toxic if swallowed. A strong purgative. It changes the urine color to red. Repeated swallowing causes kidney damage. Believed to be beneficial and healing on external skin burns.

Synonyms: ✲ LALOI ✲ SABILA ✲ SEMPERVIVUM ✲ SINKLE BIBLE ✲ STAR CACTUS ✲ ZABILA ✲ ZAVILA

ALOIN

Products: In medicines such as OTC (over-the-counter) laxatives.

Use: As a purgative.

Health Effects: A natural product with no known toxicity. Derived from *aloe vera* plant.

Synonyms: ✲ BARBALOIN

ALPHA HYDROXY ACIDS

Products: Derived from fruit, milk, and natural sources. In skin and anti-wrinkle products.

Use: To make wrinkles less noticeable, and help skin slough off dead upper-layer cells.

Health Effects: Slightly stinging upon first application. A fruit-based acid safe for pregnant women and all skin types

Synonyms: ✲ AHA

ALPHA TOCOPHEROL

Products: In baby lotions, hair cosmetic products, and deodorant products.

Use: As an antioxidant (slows reaction with oxygen) to prevent oils from becoming rancid; as a nutrient.

Health Effects: Harmless when used for intended purposes.

Synonyms: ✲ VITAMIN E

ALUMINA TRIHYDRATE

Products: In glazes, ceramic coatings, paper coatings, and cosmetics. Glass

Use: Filler and flame retardant.

and ceramic hobby production, paper correction liquids, mattresses, and makeup products.

Health Effects: Swallowing causes fever and gastrointestinal (stomach) effects. Can cause mutagenic effects (can change inherited characteristics).

Synonyms: CAS: 21645-51-2 ✲ ALUMIGEL ✲ ALUMINA HYDRATE ✲ ALUMINIC ACID ✲ ALUMINUM HYDRATE ✲ ALUMINUM HYDROXIDE GEL ✲ ALUMINUM OXIDE HYDRATE ✲ AMPHOJEL ✲ HIGILITE ✲ LIQUIGEL

ALUMINUM

Products: Powder in paints and protective coatings. In foil packaging, tubes for toothpastes, ointment tubes, cooking ware, and furniture.

Use: In paints, for corrosion resistance, and packaging.

Health Effects: Hobbyists should be aware that fine powder forms flammable and explosive mixtures in air. There is some suspicion, but no scientific evidence, that aluminum cookware, foil, or even antacid tablets containing aluminum, cause Alzheimer's disease.

Synonyms: CAS: 7429-90-5 ✲ ALUMINUM DEHYDRATED ✲ ALUMINUM FLAKE ✲ ALUMINUM POWDER ✲ METANA ALUMINUM PASTE

ALUMINUM AMMONIUM SULFATE

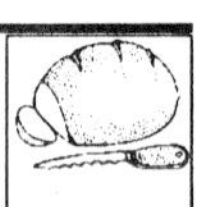

Products: A baking powder ingredient.

Use: As a buffer, neutralizing agent, and mild astringent.

Health Effects: FDA states GRAS (generally regarded as safe). Irritating if breathed or swallowed in concentrated, pure amounts.

Synonyms: None found.

ALUMINUM CHLORIDE

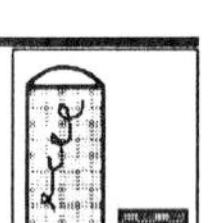

Products: In antiperspirants and deodorants.

Use: Antiperspirant.

Health Effects: Possible allergen and irritant. Moderately toxic by swallowing.

Synonyms: CAS: 7446-70-0 ✲ ALUMINUM CHLORIDE ANHYDROUS ✲ ALUMINUM CHLORIDE SOLUTION ✲ ALUMINUM TRICHLORIDE ✲ PEARSALL ✲ TRICHLOROALUMINUM

ALUMINUM DEXTRAN

Products: In packaging materials, makeup products, and various food additives.

Use: An anticaking agent, binder in cosmetic compressed powders, emulsifier, and stabilizer.

Health Effects: Must conform to FDA specifications for salts or fats or fatty acids derived from edible oils.

Synonyms: CAS: 7047-84-9 ❊ ALUMINUM MONOSTEARATE ❊ ALUMINUM STEARATE ❊ STEARIC ACID, ALUMINIUM SALT

ALUMINUM OXIDE

Products: A food additive.

Use: A dispersing agent for white coloring matter.

Health Effects: The pure chemical in gross amounts is a possible carcinogen (cancer causing substance). Breathing of fine particles can cause lung damage. Its use is limited to 2% of total ingredient. NOTE: (The gemstones ruby and sapphire are aluminum oxide colored by traces of chromium and cobalt in another form).

Synonyms: CAS: 1344-28-1 ❊ ALUMINA ❊ ALUMITE ❊ ALUNDUM ❊ ACTIVATED ALUMINUM OXIDE ❊ ALUMINUM SESQUIOXIDE.

ALUMINUM PHOSPHATE

Products: In dental cements, cosmetics, paints and varnishes.

Use: As a gelling agent, an antacid, and a cement component.

Health Effects: These solutions are corrosive to tissue.

Synonyms: CAS: 7784-30-7 ❊ ALUMINOPHOSPHORIC ACID ❊ ALUMINUM ACID PHOSPHATE ❊ ALUPHOS ❊ MONOALUMINUM PHOSPHATE ❊ PHOSPHALUGEL

ALUMINUM PHOSPHIDE

Products: In brewer's corn, grits, brewer's malt, and brewer's rice.

Use: In fumigants, insecticides, and agricultural chemicals.

Health Effects: A human poison by breathing and swallowing large amounts of pure substance. Dangerous fire risk. On EPA extremely hazardous substance list.

Synonyms: CAS: 20859-73-8 ❊ AIP ❊ AL-PHOS ❊ ALUMINUM FOSFIDE (DUTCH) ❊ ALUMINUM MONOPHOSPHIDE ❊ CELPHIDE ❊ CELPHOS ❊ DELICIA ❊ FUMITOXIN

ALUMINUM SULFATE (2:3)

Products: In animal glue, packaging materials, pickle relish, pickles, potatoes, and shrimp packs

Use: As a firming agent.

Health Effects: FDA states GRAS (generally regarded as safe) when used for intended purpose. Moderately toxic by swallowing large concentrated amounts.

Synonyms: CAS: 10043-01-3 ✣ ALUMINUM TRISULFATE ✣ CAKE ALUM ✣ DIALUMINUM SULFATE ✣ SULFURIC ACID, ALUMINUM SALT ✣ ALUM ✣ DIALUMINUM TRISULFATE

AMALGAM

Products: In dental fillings and for silvering mirrors.

Use: Binder for precious metals.

Health Effects: Mixture of mercury with other metals or alloys. Mercury is highly toxic by skin absorption or breathing of fumes.

Synonyms: ✣ MERCURY SILVER TIN ALLOY

p-AMINOAZOBENZENE

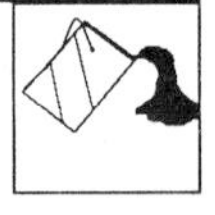

Products: Varnishes, waxes, oil stains, and insecticides.

Use: Dye.

Health Effects: Suspected carcinogen (may cause cancer).

Synonyms: CAS: 60-09-3 ✣ ANILINE YELLOW ✣ PHENYLAZOANILINE ✣ BRASILAZINA OIL YELLOW ✣ C.I. SOLVENT YELLOW ✣ ORGANOL YELLOW ✣ PARAPHENOLAZO ANILINE

p-AMINOBENZOIC ACID

Products: In suntan lotion.

Use: As an ultraviolet (UV) absorber, sunscreen.

Health Effects: Possible allergen. Moderately toxic by swallowing. Swallowing can cause nausea, vomiting, skin rash, and toxic hepatitis.

Synonyms: CAS: 150-13-0 ✣ 4-AMINOBENZOIC ACID ✣ ANTI-CHROMOTRICHIA FACTOR ✣ PABA ✣ p-CARBOXYPHENYLAMINE ✣ TRICHOCHROMOGENIC FACTOR ✣ VITAMIN H

AMMONIA ANHYDROUS

Products: In fertilizers and refrigerants.

Use: As a plant nutrient.

Health Effects: Breathing of concentrated fumes may be fatal.

Synonyms: CAS: 7664-41-7 ✣ AMMONIAC ✣ AMMONIA GAS ✣ ANHYDROUS AMMONIA ✣ SPIRIT OF HARTSHORN

AMMONIA, AROMATIC SPIRITS

Products: In inhalers.

Use: A respiratory stimulant.

Health Effects: Irritant to nose and throat.

Synonyms: ✲ 10% ALCOHOL IN AMMONIA

AMMONIA WATER

Products: In hair permanent wave solution, hair straighteners, hair coloring agents, and skin creams.

Use: Hair texture changer.

Health Effects: Toxic when breathed. Severe irritant to eyes, nose, and throat.

Synonyms: ✲ AMMONIA, LIQUID

AMMONIUM BICARBONATE

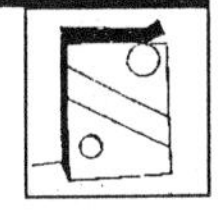

Products: In cookies, crackers, and cream-puff doughs.

Use: As a dough strengthener and fire extinguishing agent.

Health Effects: FDA states GRAS (generally regarded as safe) when used for intended purpose. Possible irritant to skin.

Synonyms: CAS: 1066-33-7 ✲ AMMONIUM ACID CARBONATE ✲ AMMONIUM HYDROGEN CARBONATE ✲ CARBONIC ACID ✲ MONOAMMONIUM CARBONATE ✲ MONOAMMONIUM SALT

AMMONIUM BORATE

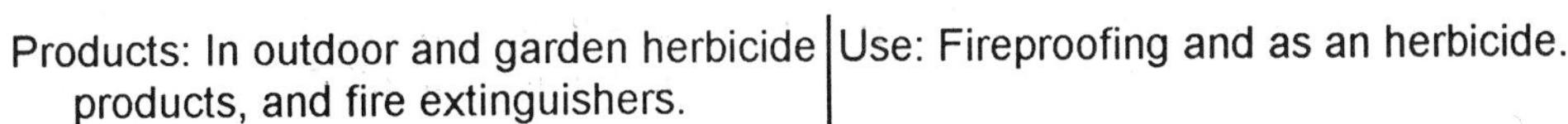

Products: In outdoor and garden herbicide products, and fire extinguishers.

Use: Fireproofing and as an herbicide.

Health Effects: Package directions must be followed carefully.

Synonyms: ✲ AMMONIUM BIBORATE

AMMONIUM CARBAMATE

Products: In garden and agricultural products, hair waving solutions, hair products, and baking powders.

Use: As a fertilizer, fire extinguisher, additive.

Health Effects: Produces irritating fumes to eyes, nose, and lungs. Can cause skin irritation.

Synonyms: CAS: 1111-78-0 ✲ CARBAMIC ACID ✲ AMMONIUM AMINOFORMATE ✲ MONOAMMONIUM SALT ✲ CARBAMIC SALT

AMMONIUM CARBONATE

Products: In baked goods, baking powder, caramel, gelatins, puddings, and wine.

Use: As a buffer, leavening agent, neutralizing agent, and yeast nutrient.

Health Effects: FDA states GRAS (generally regarded as safe) when used for intended purposes.

Synonyms: CAS: 506-87-6 ✲ CARBONIC ACID, AMMONIUM SALT ✲ DIAMMONIUM CARBONATE ✲ CRYSTAL AMMONIA ✲ AMMONIUM SESQUICARBONATE ✲ HARTSHORN

AMMONIUM CHLORIDE

Products: In baked goods, condiments, relishes, and washing powders.

Use: As a dough conditioner, dough strengthener, flavor enhancer, leavening agent, and yeast food.

Health Effects: FDA states GRAS (generally regarded as safe) when used for intended purposes. Pure chemical is moderately toxic in gross amounts. A severe eye irritant.

Synonyms: CAS: 12125-02-9 ✲ AMMONIUMCHLORID (GERMAN) ✲ AMMONIUM MURIATE ✲ CHLORID AMONNY (CZECH) ✲ SAL AMMONIA ✲ SAL AMMONIAC

AMMONIUM FLUORIDE

Products: In antiseptics, wood preservatives, moth-proofing, and glass-etching.

Use: As a preservative.

Health Effects: Corrosive to tissue.

Synonyms: CAS: 12125-01-8 ✲ NEUTRAL AMMONIUM FLUORIDE

AMMONIUM HYDROXIDE

Products: In baked goods, caramel, cheese, fruits (processed), and puddings.

Use: As a leavening agent, surface-finishing agent, ammonia soaps, fireproofing woods, detergents, household cleaners, and hair dyes.

Health Effects: FDA states GRAS (generally regarded as safe) when used for intended purposes. A poison and severe irritant to mouth and throat when large amounts of pure chemical are swallowed.

Synonyms: CAS 1336-21-6 ❋ AMMONIA AQUEOUS ❋ AMMONIA SOLLUTION ❋ AQUA AMMONIA

AMMONIUM NITRATE

Products: In fertilizer, herbicide, and insecticide. The freezing mixture ingredient in picnic and drink coolers.

Use: For garden and agricultural chemical products.

Health Effects: May explode under confinement and high temperatures, but not readily detonated.

Synonyms: CAS: 6484-52-2 ❋ NITRIC ACID, AMMONIUM SALT ❋ NORWAY SALTPETER

AMMONIUM PHOSPHATE

Products: Flameproofing of wood, paper, and textiles. Vegetation coating to retard forest fires. Added to matches and candles to prevent afterglow and smoking. In ammoniated dentifrices.

Use: In fireproofing and in dental products.

Health Effects: Harmless when used for intended purposes.

Synonyms: ❋ DIAMMONIUM HYDROGEN PHOSPHATE ❋ DIAMMONIUM PHOSPHATE ❋ DAP

AMYL ACETATE

Products: In nail polish, polish remover, leather polish, shoe polish, perfume odorant, and food.

Use: As solvent or flavoring agent.

Health Effects: Human systemic (entire body) effects by breathing. Mildly toxic. Skin and eye irritation. Flammable, dangerous fire hazard.

Synonyms: CAS: 628-63-7 ❋ ACETIC ACID, AMYL ESTER ❋ AMYL ACETIC ESTER ❋ PEAR OIL ❋ BANANA OIL ❋ BIRNENOEL ❋ PENT-ACETATE ❋ PENTYL ACETATE

AMYL FORMATE

Products: In films, coatings, leather, and perfume.

Use: As solvent, odorant, and flavoring.

Health Effects: Flammable, dangerous fire risk. Toxic by swallowing and breathing. Skin irritant.

Synonyms: CAS: 638-49-3 ❋ PENTYL FORMATE

AMYL PROPIONATE

Products: In perfumes, lacquers, and flavors.

Use: An ingredient to affect the taste or smell.

Health Effects: Concentrated chemical is combustible. Fire hazard.

Synonyms: ❖ ISOAMYL PROPIONATE

ANETHOLE

Products: In perfume, anise flavor for dentifrice, and licorice candy.

Use: As a flavoring agent.

Health Effects: Pure substance is a possible irritant to mouth, nose, and throat in excessive amounts. Possible skin allergen.

Synonyms: ❖ ANISE OIL ❖ ANISE CAMPHOR ❖ p-PROPENYLANISOLE ❖ p-METHOXYPROPENYLBENZENE

p-ANISALDEHYDE

Products: In perfumery; an intermediate for antihistamines.

Use: As a flavoring agent and in pharmaceuticals.

Health Effects: Moderately toxic by swallowing. A skin irritant. A possible mutagen (changes inherited characteristics).

Synonyms: CAS: 123-11-5 ❖ AUBEPINE ❖ p-ANISIC ALDEHYDE ❖ ANISIC ALDEHYDE ❖ p-METHOXYBENZALDEHYDE.

ANISE OIL

Products: In bakery products, beverages (alcoholic), beverages (nonalcoholic), candy, chewing gum, confections, ice cream, liquors, meat, pastries (sweet), and soups.

Use: As a flavoring agent.

Health Effects: FDA states GRAS (generally regarded as safe) when used at moderate levels to accomplish the intended results. Moderately toxic by swallowing large amounts of concentrated oil. A weak sensitizer (may cause skin irritation after second exposure). May cause contact dermatitis (skin rash and irritation).

Synonyms: CAS: 8007-70-3 ❖ ANISEED OIL ❖ ANIS OEL (GERMAN) ❖ OIL of ANISE ❖ STAR ANISE OIL

ANISIC ALCOHOL

Products: In perfume for light floral odors or flavoring.

Use: An ingredient used to affect the smell of final product.

Health Effects: A possible allergen. Moderately toxic by swallowing. A skin irritant.

Synonyms: CAS:105-13-5 ✲ p-METHOXYBEMZYL ALCOHOL ✲ ANISYL ALCOHOL ✲ ANISE ALCOHOL ✲ p-ANISOL ALCOHOL

ANTHRALIN

Products: In medications for the treatment of psoriasis on extremities.

Use: As an ointment; not for severe cases.

Health Effects: Very irritating. Not for use on scalp or near eyes.

Synonyms: ✲ 1,8,9-ANTHRACENETRIOL ✲ 1,8-DIHYDROXYANTHRANOL

ANTIBIOTIC

Products: Tyrothricin, bacitracin, polymxin, actinomycin, streptomycin, chloramphenicol, tetracycline. Certain antibiotics are used as food additives to inhibit the growth of bacteria and fungi. These are nisin, pimaricin, nystatin, tylosin.

Use: A chemical substance produced by microorganisms that has the ability to inhibit the growth of other microorganisms or destroy them.

Health Effects: A possible allergen. Overuse of antibiotics can lead to the development of resistant strains of antibiotics. Some antibiotics enter humans via the food chain. Cattle which has been treated with antibiotic medication and slaughtered too early can reach the consumer. The same can happen with milk cows which have had udder infections and were treated with antibiotics.

Synonyms: As above.

ANTIMONY TRISULFIDE

Products: In matches and camouflage paints.

Use: As a coloring agent.

Health Effects: Blood and gastrointestinal (stomach and digestive) effects by breathing. Possible carcinogen (causes cancer).

Synonyms: CAS: 1345-04-6 ✣ ANTIMONOUS SULFIDE ✣ ANTIMONY GLANCE ✣ ANTIMONY ORANGE ✣ ANTIMONY SULFIDE ✣ LYMPHOSCAN ✣ NEEDLE ANTIMONY

ANTISEPTIC

Products: Alcohol, boric acid, barates, acriflavine, menthol, hydrogen peroxide, hypochlorites, iodine, mercuric chloride, phenol, hexachlorophene, and quaternary ammonium compounds.

Use: A substance that retards or stops the growth of microorganisms.

Health Effects: Could possibly be corrosive or toxic.

Synonyms: ✣ ANTIBACTERIAL ✣ ANTIMICROBIAL

ARABIC GUM

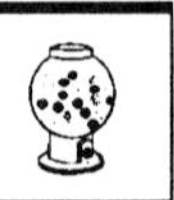

Products: In adhesives, cosmetics, thickening agent, binding agent in tablets, stabilizer in food products. Beverage bases, beverages, candy (hard), candy (soft), chewing gum, confections, cough drops, dairy product analogs, fats, fillings, frostings, gelatins, nut products, nuts, oil, puddings, confection products (quiescently frozen), and snack foods.

Use: As emulsifier, flavoring agent, formulation aid, humectant (attracts moisture), stabilizer, and surface-finishing agent.

Health Effects: FDA limits use in food products. Pure chemical is mildly toxic by swallowing large amounts. Breathing or swallowing has produced hives, eczema, and edema (swelling). Animal reproductive (infertility, or sterility, or birth defects) effects. A severe eye irritant. A weak allergen.

Synonyms: CAS: 9000-01-5 ✣ ACACIA ✣ ACACIA DEALBATA GUM ✣ ACACIA GUM ✣ ACACIA SENEGAL ✣ ACACIA SYRUP ✣ AUSTRALIAN GUM ✣ GUM ARABIC ✣ GUM OVALINE ✣ GUM SENEGAL ✣ INDIAN GUM ✣ SENEGAL GUM ✣ WATTLE GUM

ARSENIC TRIOXIDE

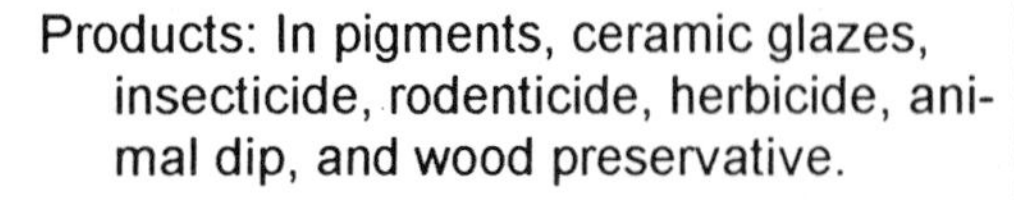

Products: In pigments, ceramic glazes, insecticide, rodenticide, herbicide, animal dip, and wood preservative.

Use: As a coloring agent, pesticide, and preservative.

Health Effects: Confirmed human carcinogen (causes cancer). Poison by swallowing. Human gastrointestinal (stomach and intestine) effects by swallowing. An animal teratogen (abnormal fetus development).

Synonyms: CAS: 1327-53-3 ✲ CRUDE ARSENIC ✲ WHITE ARSENIC ✲ ARSENIOUS ACID ✲ ARSENIOUS OXIDE ✲ ARSENOUS ANHYDRIDE

ARTEMISIA OIL

Products: In beverages (alcoholic).

Use: As a flavoring agent.

Health Effects: FDA approves use at moderate levels to accomplish the intended effect. Moderately toxic by swallowing large concentrated amounts of pure oil. An allergen. Habitual users develop "absinthism" with tremors, vertigo, vomiting, and hallucinations. May cause a contact dermatitis.

Synonyms: CAS: 8022-37-5 ✲ ABSINTHIUM ✲ ARTEMISIA OIL (WORMWOOD) ✲ OIL, ARTEMISIA

ARTIFICIAL COLORINGS

Products: Usually found in low nutritional, high sugar content foods, candy, carbonated drinks, ice sticks, and gelatins.

Use: Colorings are added when natural fruit colorings are absent. Not required to be listed on labels.

Health Effects: Avoid when possible. Problems can be anything from allergic reactions to being carcinogenic (cancer causing). See individual coloring names.

Synonyms: See specific colors.

ARTIFICIAL FLAVORINGS

Products: Frequently found in low nutritional, high sugar content foods, candy, carbonated drinks, ice sticks, gelatins, and breakfast cereals.

Use: Flavorings are added when real fruit, or natural flavor, are absent.

Health Effects: Flavorings that occur in nature are usually safe. Artificial flavorings are usually less desirable.

Synonyms: See specific flavor.

ARTIFICIAL SNOW

Products: Christmas window and display spray.

Use: Decorating.

Health Effects: Flammable. Mildly toxic by swallowing.

Synonyms: ✲ COPOLYMER OF BUTYL METHACRYLATE ✲ COPOLYMER OF ISOBUTYL METHACRYLATE

ASBESTOS

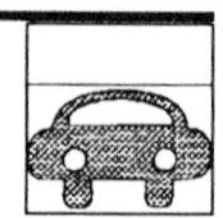

Products: In fireproof fabrics, brake linings, gaskets, roofing compositions, paint filler, and filters.

Use: For fireproofing, reinforcing agent, and fillers.

Health Effects: A carcinogen (causes cancer). Highly toxic by breathing of dust particles. Usually 4 to 7 years of exposure are required before serious lung damage could occur.

Synonyms: CAS: 1332-21-4 ✲ AMIANTHUS ✲ AMOSITE ✲ AMPHIBOLE ✲ FIBROUS GRUNERITE ✲ SERPENTINE

ASCORBIC ACID

Products: In beverages, potato flakes, breakfast foods, beef (cured), meat food products, pork (cured and fresh), sausage, and wine.

Use: As a dietary supplement, nutrient, preservative, and antioxidant (slows reaction with oxygen) to increase shelf life.

Health Effects: FDA states GRAS (generally regarded as safe) when used for intended purposes. The concentrated chemical is moderately toxic in gross amounts.

Synonyms: CAS: 50-81-7 ✲ l-ASCORBIC ACID ✲ VITAMIN C ✲ ASCORBUTINA ✲ CEVITAMIC ACID ✲ CEVITAMIN ✲ VITACIN ✲ VITAMISIN ✲ VITASCORBOL

ASPARTAME

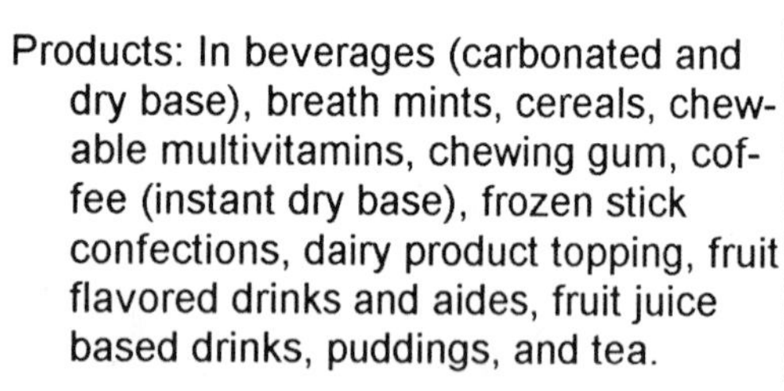

Products: In beverages (carbonated and dry base), breath mints, cereals, chewable multivitamins, chewing gum, coffee (instant dry base), frozen stick confections, dairy product topping, fruit flavored drinks and aides, fruit juice based drinks, puddings, and tea.

Use: As a flavor enhancer, sugar substitute, sweetener (approximately 71% of market).

Health Effects: FDA approves use at moderate levels to accomplish the intended results. Made from two amino acids. It is equal to sucrose in calories but it is 200 times sweeter. Therefore, the amount needed to sweeten food is negligible. Could cause allergic dermatitis. Concentrated chemical caused reproductive effects (infertility, or sterility, or birth defects) in animal studies. Aspartame does contain a chemical called phenylanine which is dangerous to people with the inherited disease phenylketonuria (PKU).

Synonyms: CAS: 22839-47-0 ✲ EQUAL ✲ CANDEREL ✲ DIPEPTIDE SWEETNER ✲ METHYL ASPARTYLPHENYLALANATE ✲ NUTRASWEET ✲ ASPARTYLPHENYLALANINE METHYL ESTER ✲ SWEET DIPEPTIDE

ASPHALT

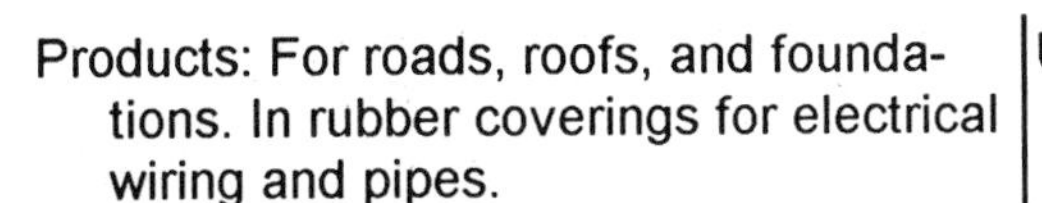
Products: For roads, roofs, and foundations. In rubber coverings for electrical wiring and pipes.

Use: As a coating and for waterproofing.

Health Effects: A suspected carcinogen (causes cancer). Some components may be carcinogens. Moderately irritating to skin. Can cause dermatitis. Fumes can be irritating to nose, eyes and lungs.

Synonyms: CAS: 8052-42-4 ❊ ASPHALTUM ❊ BITUMEN ❊ MINERAL PITCH ❊ PETROLEUM PITCH ❊ ROAD ASPHALT ❊ ROAD TAR

ASPIRIN

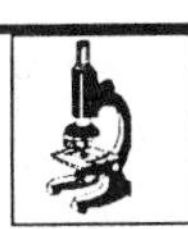

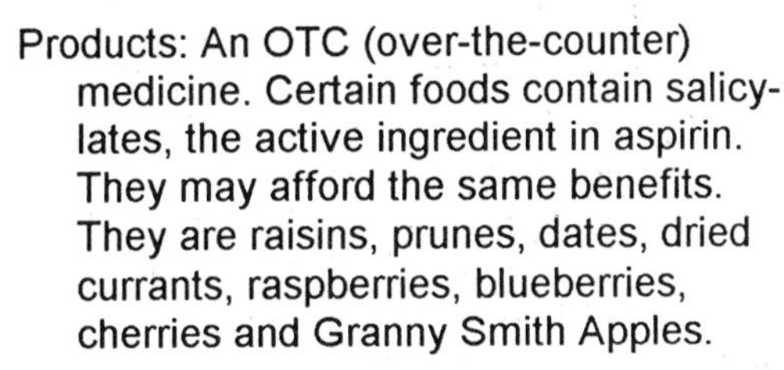
Products: An OTC (over-the-counter) medicine. Certain foods contain salicylates, the active ingredient in aspirin. They may afford the same benefits. They are raisins, prunes, dates, dried currants, raspberries, blueberries, cherries and Granny Smith Apples.

Use: An analgesic, anti-inflammatory, antipyretic (to reduce fever).

Health Effects: An allergen. May cause bleeding. A 10 g dose may be fatal. Sometimes prescribed to prevent heart attacks because of anticoagulant action. Recent studies indicate it my be a colon cancer preventative. It should be avoided before surgical procedure to avoid excessive bleeding. It should not be taken by children because of Reyes syndrome implication.

Synonyms: ❊ ACETYLSALICYLIC ACID ❊ o-ACETOXYBENZOIC ACID

AZINPHOS METHYL

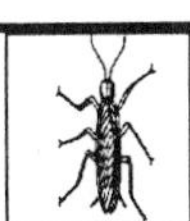

Products: For fruit, citrus pulp, and soybean oil. Use limited to 5 ppm (parts per million) by FDA.

Use: As an insecticide.

Health Effects: Poison by breathing, swallowing large amounts, and skin contact. An animal teratogen (abnormal fetus development). Other animal reproductive effects (infertility, or sterility, or birth defects). A human mutagen (changes inherited characteristics). A possible carcinogen (may cause cancer).

Synonyms: CAS: 86-50-0 ❊ AZINPHOS METHYL ❊ BENZOTRIAZINE derivative of a METHYL DITHIOPHOSPHATE ❊ BENZOTRIAZINEDITHIOPHOSPHORIC ACID DIMETHOXY ESTER ❊ CARFENE ❊ COTNION METHYL ❊ CRYSTHYON ❊ DBD ❊ GOTHNION ❊ GUSATHION

B

Just because something is natural does not mean it is necessarily healthy and good. Poison Ivy and Salmonella are natural.

BAKER'S YEAST EXTRACT

Products: In cheese spread flavorings, frozen desserts, salad dressings, cheese flavored snack dips, soups, sour cream and wines.

Use: As an emulsifier, flavoring agent, nutrient supplement, stabilizer, and thickener yeast food.

Health Effects: FDA states GRAS (generally regarded as safe) when used for intended purpose. Limited to 5% in salad dressing.

Synonyms: ✲ AUTOLYZED YEAST EXTRACT ✲ BAKER'S YEAST GLYCAN

BAKING SODA

Products: In food, mouth washes, antacids, skin powders, and bath salts.

Use: As a leavening agent or to adjust acidity in foods. In laundry or refrigerator as deodorizer.

Health Effects: In rare cases there have been reports of people taking baking soda internally for indigestion after large meals and experiencing stomach rupture. This occurs because baking soda releases carbon dioxide and the stomach is too full, or there is damage as a result of ulcers or obstruction. It has high sodium content which can contribute to high blood pressure, heart disease, and kidney problems. GRAS (generally regarded as safe).

Synonyms: ✲ SODIUM BICARBONATE

BALSAM of PERU

Products: Derived from evergreen trees or shrubs. In chocolate manufacturing, expectorants, cough syrups, shampoo fragrances, and hair conditioners.

Use: As a flavoring and fragrance.

Health Effects: An allergen. Combustible when heated.

Synonyms: ✲ BALSAM PERU OIL ✲ PERUVIAN BALSAM

BARIUM CHLORIDE

Products: In lubricating oils and textile dyes.

Use: As an additive or pigment.

Health Effects: A poison when swallowed. Some toxicity upon breathing.

Synonyms: CAS: 10361-37-2 ✲ BARIUM DICHLORIDE

BARIUM THIOSULFATE

Products: In luminous paints, matches, and varnishes.

Use: As pigments, glazes, and protective coatings.

Health Effects: Flammable. Toxic by swallowing and breathing.

Synonyms: ✲ BARIUM HYPOSULFITE

BATTERY ACID

Products: Electrolyte acid.

Use: Storage battery.

Health Effects: Corrosive to skin and tissue.

Synonyms: ✲ SULFURIC ACID

BAYOIL

Products: In bay rum, meats, soups, and stews.

Use: For fragrances and flavoring. An ingredient that affects the taste or smell of final product.

Health Effects: FDA states GRAS (generally regarded as safe) when used for intended purposes. It is moderately toxic by swallowing large amounts of the pure substance.

Synonyms: ✲ BAYLEAF OIL ✲ LAUREL LEAF OIL ✲ MYRCIA OIL ✲ OIL OF BAY

BEESWAX

Products: In candy (hard and soft), candy glaze, creams, ear plugs, lipstick, and church candles.

Use: As a flavoring; in polishes, waxes, adhesives, and textile sizing.

Health Effects: A mild allergen to susceptible individuals.

Synonyms: CAS: 8012-89-3 ✲ BEESWAX, WHITE ✲ BEESWAX, YELLOW

BENTONITE

Products: In foods such as fruit flavorings and liquors. In ice cream, baked goods, and chewing gum. In cosmetics, facial makeup, facial masks, to clarify and stabilize wines.

Use: As a colorant, pigment, stabilizer, and thickener.

Health Effects: FDA states GRAS (generally regarded as safe) when used for intended purposes.

Synonyms: CAS: 1302-78-9 ❖ ALBAGEL PREMIUM USP 4444 ❖ HI-JEL ❖ MAGBOND ❖ MONTMORILLONITE ❖ PANTHER CREEK BENTONITE ❖ SOUTHERN BENTONITE ❖ TIXOTON ❖ VOLCLAY ❖ WILKINITE

BENZALDEHYDE

Products: In perfume, soaps, cosmetic oils, and dyes.

Use: As a flavoring and odorant.

Health Effects: GRAS (generally regarded as safe) when used at moderate levels to accomplish the desired result. An allergen. Poison by swallowing. Has very mild local anesthetic properties. A skin irritant. Causes convulsions in large doses.

Synonyms: CAS: 100-52-7 ❖ ARTIFICIAL ALMOND OIL ❖ BENZENECARBALDEHYDE ❖ BENZENECARBONAL ❖ BENZOIC ALDEHYDE

BENZENE

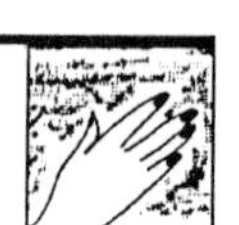

Products: In cosmetics, perfumes, nail polish remover, airplane glues, lacquers, dry cleaning products, paint, spot remover, varnish, stain, and sealant.

Use: As a solvent, coating, and various other uses.

Health Effects: A confirmed human carcinogen (causes cancer) produces myeloid leukemia, Hodgkin's disease, and lymphomas by breathing. A human poison by breathing. Moderately toxic by swallowing. A severe eye and skin irritant. Effects on the body by swallowing and breathing are blood changes, and increased body temperature. Effects are cumulative. A human mutagen (changes inherited characteristics). A narcotic. Benzene can penetrate the skin and cause poisoning. Products containing 5% or more by weight must be labeled "Danger Vapor Harmful, Poison" and the skull and crossbones symbol. If it contains 10% or more it must also state "Harmful or fatal if swallowed". "Call physician immediately". This is the highest volume chemical produced in the U.S.

Synonyms: CAS: 71-43-2 ❖ BENZOL ❖ BENZOLE ❖ BENZOLENE ❖ BICARBURET OF HYDROGEN ❖ CARBON OIL ❖ COAL NAPHTHA ❖ MOTOR BENZOL ❖ NITRATION BENZENE ❖ PHENE ❖ PYROBENZOL ❖ PYROBENZOLE ❖ PHENYL HYDRIDE

BENZENEACETALDEHYDE

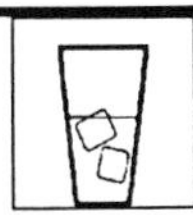

Products: In bakery products, beverages, chewing gum, confections, gelatins, ice cream, maraschino cherries, and puddings.

Use: As a flavoring.

Health Effects: It is moderately toxic by swallowing large amounts of concentrated substance. A skin irritant.

Synonyms: CAS: 122-78-1 ❊ HYACINTHIN ❊ PAA ❊ PHENYLACETALDEHYDE ❊ PHENYLACETIC ALDEHYDE ❊ PHENYLETHANAL ❊ α-TOLUALDEHYDE ❊ α-TOLUIC ALDEHYDE

BENZENE HEXACHLORIDE

Products: As a pesticide.

Use: Various.

Health Effects: A confirmed carcinogen that caused cancerous tumors by swallowing and by skin contact in animals. Poison by swallowing and skin contact. Effects on the human body by breathing are headache, nausea or vomiting, and fever. Implicated in aplastic anemia. Lindane is more toxic than DDT or dieldrin.

Synonyms: CAS: 608-73-1 ❊ BHC ❊ DBH ❊ GAMMEXANE ❊ HCCH ❊ HEXA ❊ HEXACHLOR ❊ HEXACHLORAN ❊ HEXACHLOROCYCLOHEXANE ❊ 1,2,3,4,5,6-HEXACHLOROCYCLOHEXANE ❊ HEXYLAN ❊ LINDANE

BENZOCAINE

Products: In medicine (local anesthetic) and in suntan preparations.

Use: To numb or deaden skin to prevent pain.

Health Effects: Toxic by swallowing.

Synonyms: CAS: 51-05-8 ❊ ANESTHESOL ❊ PROCAINE HYDROCHLORIDE ❊ ETHYL-p-AMINOBENZOATE HYDROCHLORIDE

BENZOIC ACID

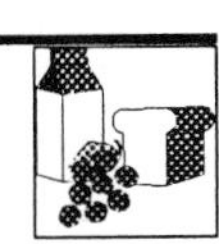

Products: Cinnamon (naturally occurring), cloves (ripe, naturally occurring), cranberries (naturally occurring), plums (naturally occurring), and prunes (naturally occurring). In food preservatives, tobacco seasoning, flavors, perfumes, and toothpastes.

Use: As an antimicrobial (antiseptic) agent, flavoring agent, and preservative.

Health Effects: The pure acid is moderately toxic by swallowing gross amounts. Effects on the body by breathing are dyspnea (shortness of breath) and skin allergy. A severe eye irritant and skin irritant. GRAS (generally regarded as safe) with a limitation of 0.1 percent in foods.

Synonyms: CAS: 65-85-0 ❊ BENZENECARBOXYLIC ACID ❊ BENZENEFORMIC ACID ❊ BENZENEMETHANOIC ACID ❊ BENZOATE ❊ CARBOXYBENZENE ❊ DRACYLIC ACID ❊ PHENYL CARBOXYLIC ACID ❊ PHENYLFORMIC ACID ❊ RETARDEX ❊ SALVO

BENOMYL

Products: Pesticide.

Use: On raisin and tomato products.

Health Effects: Poison by swallowing. Mildly toxic by breathing. An animal teratogen (abnormal fetus development). Caused reproductive effects in animal studies (infertility, or sterility, or birth defects). A human mutagen (changes inherited characteristics). A skin irritant. EPA was ordered by courts to revoke approval as food additive in 1992.

Synonyms: CAS: 17804-35-2 ❊ BNM ❊ BENLATE ❊ FUNDASOL

BENZOTRIFLUORIDE

Products: Solvents and insecticides.

Use: Pesticides.

Health Effects: Highly toxic by breathing. Flammable, dangerous fire risk.

Synonyms: CAS: 98-08-8 ❊ BENZENYL FLUORIDE ❊ BENZYLIDYNE FLOURIDE ❊ PHENYLFLUOROFORM ❊ TRIFLOUROMETHYL BENZENE ❊ TOLUENE TRIFLUORIDE

BENZOYL PEROXIDE

Products: In cheese (asiago fresh), cheese (asiago medium), cheese (asiago old), cheese (asiago soft), cheese (blue), cheese (caciocavallo siciliano), cheese (emmentaler), cheese (gorgonzola), cheese (parmesan), cheese (provolone), cheese (reggiano), cheese (romano), cheese (Swiss), margarine, sausage casings, shortening, and whey (annatto-colored). Also in acne cream.

Use: As bleaching agent for flour, fats, oils and waxes. Anti-acne pharmaceuticals and cosmetics, cheese production, and in the embossing of vinyl flooring.

Health Effects: A poison by swallowing large amounts of concentrated chemical. Can cause dermatitis. An allergen and eye irritant. A mutagen (changes inherited characteristics). A possible carcinogen (causes cancer) that caused tumor growth in animals. GRAS (generally regarded as safe).

Synonyms: CAS: 94-36-0 ❊ ACETOXYL ❊ ACNEGEL ❊ BENOXYL ❊ BENZAC ❊ BENZOIC ACID, PEROXIDE ❊ BENZOPEROXIDE ❊ BENZOYL ❊ BENZOYL SUPEROXIDE ❊ CLEARASIL BENZOYL PEROXIDE LOTION ❊ CLEARASIL BP ACNE TREATMENT ❊ CUTICURA ACNE CREAM ❊ DEBROXIDE ❊ DRY AND CLEAR ❊ EPI-CLEAR ❊ FOSTEX ❊ OXY-5 ❊ OXY-10 ❊ OXYLITE ❊ OXY WASH

BENZYL ACETATE

Products: In artificial jasmine, perfumes, soap perfume, lacquers, polishes, inks, and varnish remover.

Use: As a flavoring, solvent, and odorant.

Health Effects: Highly toxic by breathing. Skin irritant.

Synonyms: CAS: 140-11-4 ✲ ACETIC ACID BENZYL ESTER ✲ ACETIC ACID PHENYLMETHYL ESTER ✲ BENZYL ETHANOATE

BENZYL ALCOHOL

Products: In perfumes and flavors, bacteriostat, cosmetics, ointments, ballpoint pen ink, stencil ink, and hair dye preservative.

Use: As a flavoring, odorant, solvent, preservative, and pharmaceutical.

Health Effects: Highly toxic. Poison by swallowing. Moderately toxic by breathing and skin contact. A moderate skin and severe eye irritant.

Synonyms: CAS: 100-51-6 ✲ BENZAL ALCOHOL ✲ BENZENECARBINOL ✲ BENZENEMETHANOL ✲ PHENOCARBINOL ✲ PHENYLMETHANOL ✲ PHENYL-METHYL ALCOHOL

BENZYL BENZOATE

Products: In perfume musk, external medicines, plasticizer in nail polish, and miticide.

Use: As a fixative, solvent, flavor, and plasticizer.

Health Effects: Possible allergen. Irritant to eyes and skin.

Synonyms: CAS: 120-51-4 ✲ ASCABIN ✲ ASCABIOL ✲ BENYLATE ✲ BENZYL BENZENECARBOXYLATE ✲ BENZYL BENZFORMATE ✲ NOVOSCABIN ✲ VENZONATE

o-BENZYL-p-CHLOROPHENOL

Products: In soaps.

Use: As a disinfectant

Health Effects: Highly toxic. An irritant. Combustible.

Synonyms: ✲ CHLOROPHENE ✲ SANTOPHEN ✲ SEPTIPHENE ✲ 4-CHLORO-α-PHENYL-o-CRESOL

BENZYL CINNAMATE

Products: In fruit scented perfumes and flavors.

Use: As a flavoring and odorant. An ingredient that affects the taste of smell of the final product.

Health Effects: In large amounts the pure substance is moderately toxic by swallowing. A mild allergen and skin irritant.

Synonyms: CAS: 103-41-3 ✲ BENZYL ALCOHOL CINNAMIC ESTER ✲ BENZYL γ-PHENYLACRYLATE ✲ CINNAMEIN ✲ trans-CINNAMIC ACID BENZYL ESTER ✲ 3-PHENYL-2-PROPENOIC ACID PHENYLMETHYL ESTER (9CI)

---

p-BENZYLPHENOL

Products: In soaps and cleansers.

Use: As an antiseptic and germicide.

Health Effects: Toxic by swallowing. Possible allergen.

Synonyms: ✲ 4-HYDROXY DIPHENYLMETHANE

BENZYL SALICYLATE

Products: In perfumes for cosmetics, musk perfume, sun-screening products, and soaps.

Use: As a fixative, solvent, and odorant.

Health Effects: A possible allergen. Skin irritant. Moderately toxic by swallowing.

Synonyms: CAS: 118-58-1 ✲ BENZYL-o-HYDROXY BENZOATE

BENZYL THIOCYANATE

Products: In insecticides.

Use: As a pesticide.

Health Effects: A strong irritant to skin and tissue. Moderate fire hazard. Avoid use around open flame or campfire.

Synonyms: CAS: 3012-37-1 ✲ BENZYL MUSTARD OIL ✲ PHENYLMETHYL ESTER THIOCYANIC ACID ✲ TROPEOLIN

BENZYL THIOL

Products: In beverages, baked goods, candy, and ice cream.

Use: As an odorant and flavoring.

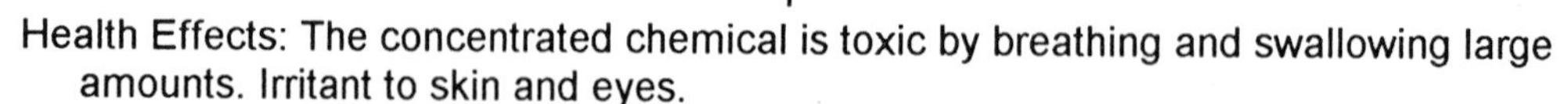

Health Effects: The concentrated chemical is toxic by breathing and swallowing large amounts. Irritant to skin and eyes.

Synonyms: ✲ BENZYL MERCAPTAN ✲ α-TOLUENETHIOL

BERGAMOT OIL rectified

Products: In bakery products, beverages (alcoholic), chewing gum, confections, gelatin desserts, ice cream, and

Use: An ingredient that affects taste or smell of product.

puddings. In perfumery for hair tonic, oils, and dressings.

Health Effects: It is mildly toxic by swallowing large amounts of pure substance. A mild skin irritant and allergen. GRAS (generally regarded as safe).

Synonyms: CAS: 8007-75-8 ✲ BERGAMOTTE OEL (GERMAN) ✲ OIL of BERGAMOT, coldpressed ✲ OIL of BERGAMOT, rectified

BETA CAROTENE

Products: In orange beverages, desserts, cheese, ice cream, margarine, shortening, butter, non-dairy whiteners, and cosmetics.

Use: As a coloring and nutrient.

Health Effects: Natural yellow coloring found in carrots, vegetables and some animal tissue. GRAS (generally regarded as safe).

Synonyms: ✲ VITAMIN A DERIVATIVE. ✲ B-CAROTENE

BOVINE GROWTH HORMONE

Products: In livestock and cattle.

Use: Increases size of animal.

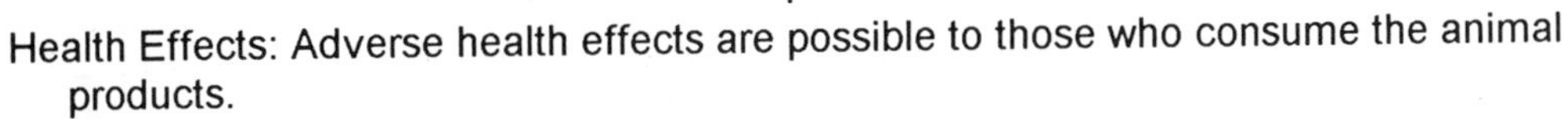

Health Effects: Adverse health effects are possible to those who consume the animal products.

Synonyms: ✲ BGH

BHA

Products: In meats, cereals, chewing gum, desserts, shortening, dry fruit, margarine, pizza toppings, potato products, poultry, rice and sausage.

Use: As preservative and antioxidant (slows down spoiling of fats due to oxidation).

Health Effects: In large amounts it is moderately toxic by swallowing concentrated chemical. Caused cancer and tumors in animal studies. Animal reproductive (infertility, or sterility, or birth defects) and mutagenic effects (changes inherited characteristics). Use is limited by FDA regulations.

Synonyms: CAS: 25013-16-5 ✲ ANTRANCINE ✲ BUTYLATED HYDROXYANISOLE ✲ EMBANOX ✲ NIPANTIOX ✲ PREMERGE PLUS ✲ SUSTANE ✲ VERTAC

BHT (food grade)

Products: In beef patties (fresh), beef patties (pregrilled), beet sugar, cereals

Use: As an antioxidant (slows down spoiling due to oxygen), and a

(dry breakfast), chewing gum, emulsion stabilizers for shortening, fats (rendered animal), margarine, meat (dried), meatballs (cooked or raw), oleomargarine, pizza toppings (cooked or raw), pork, potato flakes, potato granules, potato shreds (dehydrated), poultry, sausage (brown and serve), sausage (dry), sausage (fresh Italian), sweet potato flakes, and yeast.

preservative.

Health Effects: Moderately toxic by swallowing great amounts of pure substance. An animal carcinogen (causes cancer). Animal reproductive (infertility, or sterility, or birth defects) effects. A human skin irritant. A skin and eye irritant in animals. Use is permitted by the USDA with specific limitations as to amounts of additive.

Synonyms: CAS: 128-37-0 ✲ ADVASTAB 401 ✲ AGIDOL ✲ ANTIOXIDANT DBPC ✲ ANTIOXIDANT 29 ✲ 2,6-BIS(1,1-DIMETHYLETHYL)-4-METHYLPHENOL ✲ BUTYLATED HYDROXYTOLUENE ✲ BUTYLHYDROXYTOLUENE ✲ DIBUTYLATED HYDROXYTOLUENE ✲ 2,6-DI-tert-BUTYL-p-CRESOL ✲ 3,5-DI-tert-BUTYL-4-HYDROXYTOLUENE ✲ 2,6-DI-tert-BUTYL-4-METHYLPHENOL ✲ IMPRUVOL ✲ IONOL ✲ SUSTANE ✲ TOPANOL

BISMUTH

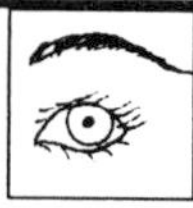

Products: In pharmaceuticals, eye shadow, lipstick, bleaching skin creams, and hair dye products.

Use: In medicines and cosmetics.

Health Effects: A possible allergen. Poisonous to humans.

Synonyms: ✲ METALLIC ELEMENT OF ATOMIC NUMBER 83 ✲ BISMUTH-209

BISODIUM TARTRATE

Products: In fats, jams, jellies, margarine, meat products, oil, oleomargarine, and sausage casings.

Use: As an emulsifier (stabilizes and maintains mixes), sequestrant (binds ingredients that affect the final products appearance, flavor, or texture).

Health Effects: The pure chemical is moderately toxic by swallowing gross amounts. GRAS (generally regarded as safe).

Synonyms: CAS: 868-18-8 ✲ DISODIUM TARTRATE ✲ DISODIUM L(+)-TARTRATE ✲ SODIUM TARTRATE (FCC) ✲ SODIUM L(+)-TARTRATE

BIS(TRIBUTYL TIN)OXIDE

Products: In underwater paints, antifouling paints, and boat bottom paints.

Use: As fungicide and bactericide in paints.

Health Effects: Toxic by swallowing and breathing. Moderately toxic by skin contact. A severe eye irritant. Animal reproduction effects (infertility, or sterility, or birth defects).

Synonyms: CAS: 56-35-9 ✲ BIOMET TBTO ✲ BTO ✲ BUTINOX ✲ HEXABUTYLDITIN ✲ TRI-n-BUTYL-STANNAE OXIDE

BITHIONOL

Products: In deodorant, germicide, fungistat, and pharmaceuticals.

Use: As a preservative and bacteriostat (kills bacteria).

Health Effects: A skin irritant. FDA states it may not be used in cosmetics. EPA lists it on its Extremely Hazardous Substances list. Possible carcinogen (may cause cancer). A poison.

Synonyms: CAS: 97-18-7 ✲ ACTAMER ✲ BIDIPHEN ✲ BITHIONOL SULFIDE ✲ BITIN ✲ NEOPELLIS ✲ LOROTHIODOL

BITTER ALMOND OIL

Products: In baked goods, beverages (alcoholic), beverages (nonalcoholic), cakes, chewing gum, confections, gelatin desserts, ice cream, maraschino cherries, pastries, and puddings.

Use: As a flavoring agent.

Health Effects: A human poison by swallowing excessive amounts of the pure substance. Moderately toxic by skin contact. A skin irritant. GRAS (generally regarded as safe).

Synonyms: CAS: 8013-76-1 ✲ ALMOND OIL BITTER ✲ OIL, BITTER ALMOND

BLACK PEPPER OIL

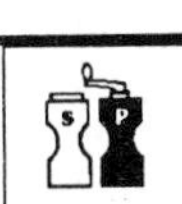

Products: On meat, salads, soups, and vegetables.

Use: As a flavoring or seasoning agent.

Health Effects: A moderate skin irritant. GRAS (generally regarded as safe) when used at a level not in excess of the amount reasonably required to accomplish the intended results.

Synonyms: CAS: 8006-82-4

BLACK POWDER

Products: Blasting powder, fireworks, fuses, igniters, or gunpowder.

Use: In time fuses for blasting and shells. In igniter and primer assemblies for propellants, fireworks, and mining.

Health Effects: Sensitive to heat. Does not detonate but is a dangerous fire and explosion hazard.

Synonyms: ❊ POTASSIUM NITRATE/CHARCOAL/SULFUR

BLEACH

Products: In clothing and textile bleach, algicide and disinfectant for pools, bactericide, deodorant, water purifier, and fungicide

Use: For whitening, purifying, and disinfecting.

Health Effects: Possibly toxic. Fire risk when in contact with organic material.

Synonyms: ❊ HYDROGEN PEROXIDE ❊ SODIUM HYPOCHLORITE ❊ SODIUM PEROXIDE ❊ SODIUM CHLORITE ❊ CALCIUM HYPOCHLORITE ❊ HYPOCHLOROUS ACID ❊ SODIUM PERBORATE ❊ DI CHLORO DIMETHYL HYDANTOIN

BLUE NUMBER 1

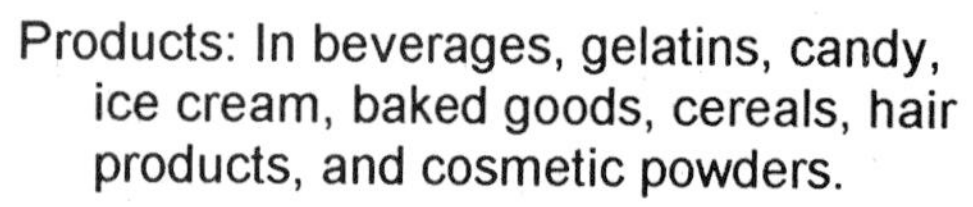

Products: In beverages, gelatins, candy, ice cream, baked goods, cereals, hair products, and cosmetic powders.

Use: As color in food, beverages, hair, and makeup products.

Health Effects: A possible allergen. In gross amounts the pure substance is a possible carcinogen (may cause cancer). World Health Organization approves use for non-food.

Synonyms: ❊ FD AND C BLUE NO. 1 ❊ FOOD DRUG AND COSMETIC BLUE 1 ❊ TRIPHENYLMETHANE

BLUE NUMBER 2

Products: In beverages, cereals, candy, soft drink powders, and jellies.

Use: As color for food.

Health Effects: The concentrated chemical is moderately toxic by swallowing excessive amounts. Possible allergen. In animal studies caused tumors and mutagenic effects (changes inherited characteristics). World Health Organization states that data is not sufficient to meet requirements acceptable for food use.

Synonyms: CAS: 860-22-0 ❊ FD AND C BLUE NO. 2 ❊ FOOD DRUG AND COSMETIC BLUE 2 ❊ ACID BLUE W ❊ ANILINE CARMINE POWDER ❊ INDIGO CARMINE ❊ INDIGOTINE

BORAX PENTAHYDRATE

Products: Weed killer, soil sterilant, and fungus controller on citrus fruit. FDA limits residue on fruit.

Use: As a defoliant and fungicide.

Health Effects: Toxic.

Synonyms: None found.

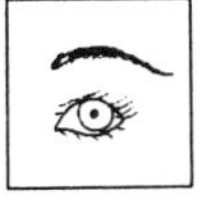

BORIC ACID

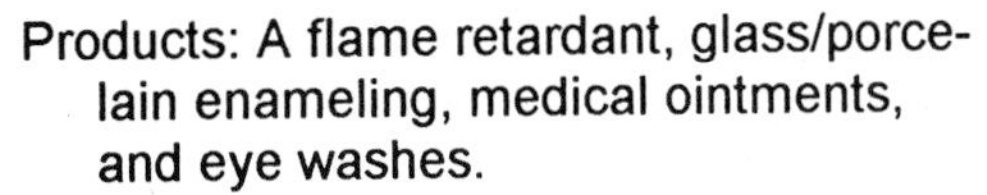

Products: A flame retardant, glass/porcelain enameling, medical ointments, and eye washes.

Use: In coatings, hobby work, and pharmaceuticals.

Health Effects: In concentrated amounts it is a poison by swallowing. Mildly toxic by skin contact. Swallowing may cause diarrhea, cramps, skin lesions, throat and mouth lesions, irregular heart beat, convulsions, and coma.

Synonyms: CAS: 10043-35-3 ✲ BORACIC ACID ✲ BOROFAX ✲ ORTHOBORIC ACID ✲ THREE ELEPHANT

BORIC OXIDE

Products: In paints and herbicides.

Use: Adds fire resistance to paints. A defoliant and pesticide.

Health Effects: Moderately toxic by swallowing. An eye and skin irritant.

Synonyms: CAS: 1303-86-2 ✲ BORIC ANHYDRIDE ✲ BORON OXIDE

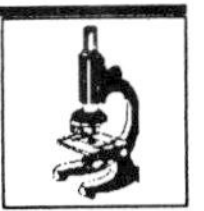

BOTULISM

Products: Usually found in foods left at room temperature.

Use: Most potent bacterial poison known.

Health Effects: Poison of toxin is produced by an organism called *Clostridium Botulinum.* Found in soil and sea water. It can form spores which can contaminate food that is improperly processed, canned, preserved or cooled. The spores can resist low cooking temperatures, then produce active organisms that in turn produce the toxins that taint the food. The toxin is one that affects the nerves of the peripheral nervous system. Double vision, dry mouth, nausea, vomiting and progressive paralysis are symptoms. Humans and most animals are susceptible. Pigs, dogs and cats are somewhat immune.

Synonyms: ✲ *CLOSTRIDIUM BOTULINUM*

BROMELIN

Products: In beer, bread, cereals (precooked), meat (raw cuts), poultry, and wine.

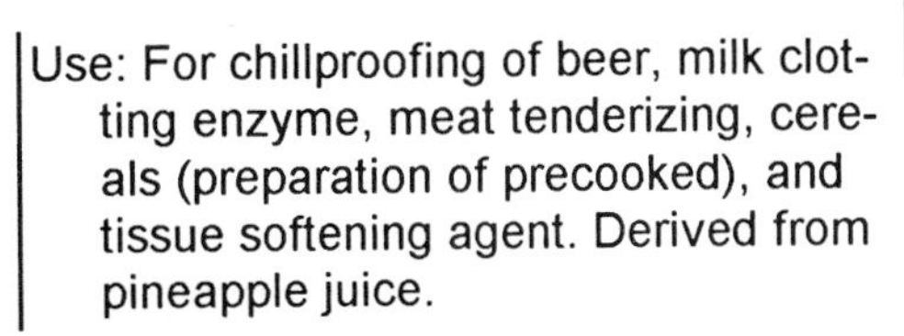

Use: For chillproofing of beer, milk clotting enzyme, meat tenderizing, cereals (preparation of precooked), and tissue softening agent. Derived from pineapple juice.

Health Effects: GRAS (generally regarded as safe).

Synonyms: CAS: 9001-00-7 ✲ ANANASE ✲ BROMELAINS ✲ EXTRANASE ✲ INFLAMEN ✲ PLANT PROTEASE CONCENTRATE ✲ TRAUMANASE

BROMINATED VEGETABLE (SOYBEAN) OIL

Products: In soft drinks (fruit flavored).

Use: As beverage stabilizer, flavoring agent, emulsifier, and clouding agent.

Health Effects: FDA approves limited use.

Synonyms: ✲ VEGETABLE (SOYBEAN) OIL, brominated ✲ BVO

BROMOACETONE

Products: For tear gas sprays, guns, aerosols, and bombs.

Use: For crowd control and protection.

Health Effects: Moderately toxic by breathing and skin contact. A lachrymator (causes eyes to water) and strong irritant.

Synonyms: CAS: 598-31-7 ✲ ACETONYL BROMIDE ✲ ACETYL METHYL BROMIDE ✲ BROMO-2-PROPANONE ✲ MONOBROMOACETONE

BROMOCHLOROMETHANE

Products: In fire extinguishers.

Use: Gas to extinguish flames.

Health Effects: A poison. Mildly toxic by swallowing and breathing. Has a narcotic effect.

Synonyms: CAS: 74-97-5 ✲ METHYLENE CHLOROBROMIDE ✲ CHLOROBROMOMETHANE ✲ HALON 1011

BROMOTRIFLUOROMETHANE

Products: For refrigerant and fire extinguishment.

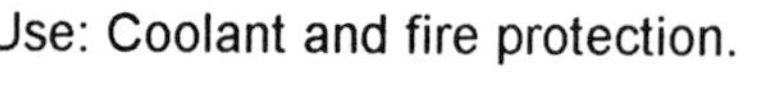

Use: Coolant and fire protection.

Health Effects: Toxic by breathing.

Synonyms: CAS: 75-63-8 ✲ TRIFLUOROBROMOMETHANE ✲ BROMOFLUOROFORM ✲ FREON 13B1 ✲ HALON 1301

BRONZE

Products: In paint, powder cosmetics, hair coloring, pearl finish eye shadow, cosmetics, and art materials.

Use: For cosmetics, hair products, and decorating.

Health Effects: Powder is flammable.

Synonyms: ✲ COPPER/TIN

BUTANE

Products: As liquid fuel. Cigarette, cigar, candle, outdoor grill, and fireplace lighters.

Use: An aerating agent, gas, and propellant.

Health Effects: Mildly toxic via breathing. Causes drowsiness. An asphyxiant (causes suffocation). Very dangerous fire hazard when exposed to heat or flame. Highly explosive when exposed to flame. To fight fire, stop flow of gas. Narcotic in high concentrations. GRAS (generally regarded as safe).

Synonyms: CAS: 106-97-8 ✲ n-BUTANE ✲ BUTANEN ✲ DIETHYL ✲ METHYLETHYLMETHANE

1,3-BUTANEDIOL

Products: In food and cosmetic products.

Use: As a preservative and to retain scent in products.

Health Effects: Mildly toxic by swallowing excessive amounts of pure substance. A skin and eye irritant.

Synonyms: CAS: 107-88-0 ✲ β-BUTYLENE GLYCOL ✲ METHYLTRIMETHYLENE GLYCOL ✲ 1,3-BUTYLENE GLYCOL

n-BUTYL ACETATE

Products: In fruit flavorings for beverages, dessert ices, ice cream, candy, gum, baked goods, and gelatins. Perfumes, nail polishes, polish remover, and lacquers.

Use: As flavoring or solvent.

Health Effects: Pure chemical is mildly toxic by breathing and swallowing large amounts. A skin and severe eye irritant. Effects by breathing are unspecified nasal and respiratory system problems. A mild allergen. High concentrations are irritating to eyes, respiratory tract, and cause narcosis (semi-unconsciousness).

Synonyms: CAS: 123-86-4 ✲ ACETIC ACID n-BUTYL ESTER ✲ BUTILE ✲ BUTYL ACETATE ✲ 1-BUTYL ACETATE ✲ BUTYL ETHANOATE

BUTOXYETHANOL

Products: In haircoloring products.

Use: As dye or solvent.

Health Effects: Poison by swallowing and skin contact. Moderately toxic via breathing: nausea, vomiting, nose tumors, headaches. Animal teratogenic (abnormal fetus development), reproductive effects (infertility, or sterility, or birth defects).

Synonyms: CAS: 111-76-2 ✼ BUTYL CELLOSOLVE ✼ BUTYL GLYCOL ✼ ETHYLENE GLYCOL MONOBUTYL ✼ POLY-SOLV EB

tert-BUTYL ACETATE

Products: Gasoline additive.

Use: As an additive and solvent.

Health Effects: Flammable, moderate fire risk. Poison by breathing and swallowing.

Synonyms: CAS: 540-88-5 ✼ ACETIC ACID tert BUTYL ESTER ✼ TLA ✼ TEXACO LEAD APPRECIATOR

n-BUTYL ALCOHOL

Products: In shampoos. A solvent for polishes. In beverages, ice desserts, creams, candy, and baked goods. In the production of fruit, liquor, and butter flavorings. Confectioneries, tablet form food supplements, and gum.

Use: As color diluent and flavoring agent.

Health Effects: Possible allergen. Concentrated chemical is moderately toxic by skin contact and swallowing. A severe eye irritation. Various respiratory system and nasal effects.

Synonyms: CAS: 71-36-3 ✼ 1-BUTANOL ✼ n-BUTANOL ✼ BUTAN-1-OL ✼ BUTANOL (DOT) ✼ BUTYL ALCOHOL ✼ BUTYL HYDROXIDE ✼ 1-HYDROXYBUTANE ✼ METHYLOLPROPANE ✼ PROPYLCARBINOL ✼ PROPYLMETHANOL

sec-BUTYL ALCOHOL

Products: Paint remover and cleaners.

Use: In solvents.

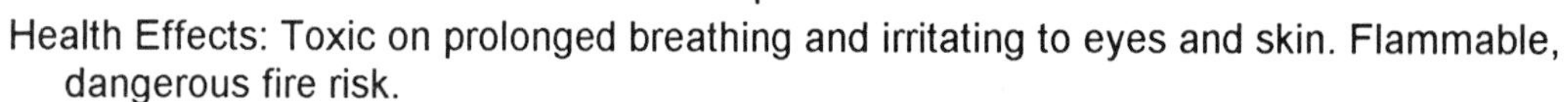

Health Effects: Toxic on prolonged breathing and irritating to eyes and skin. Flammable, dangerous fire risk.

Synonyms: CAS: 78-92-2 ✼ SBA ✼ 2-BUTANOL ✼ METHYLETHYLCARBINOL

tert-BUTYL ALCOHOL

Products: Octane booster in unleaded gasoline (EPA approved). In perfumery and as a paint remover.

Use: As a solvent, gasoline additive, and odorant.

Health Effects: Irritant to eyes and skin. Flammable, dangerous fire risk.

Synonyms: CAS: 75-65-0 ✼ 2-METHYL-2PROPANOL ✼ TRIMETHYL CARBINOL

BUTYL-p-AMINOBENZOATE

Products: Ointment for burns. An

Use: Local anesthetic and as a

ultraviolet (UV) absorber in suntan products. sunscreen.

Health Effects: Toxic by swallowing.

Synonyms: CAS: 94-25-7 ✲ p-AMINOBENZOATE ACID BUTYL ESTER ✲ BUTAMBEN

BUTYL ISOVALERATE

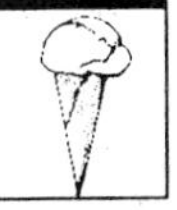

Products: In beverages, ice desserts, creams, candy, bakery goods, candy and gelatins.

Use: In chocolate and fruit flavorings.

Health Effects: It is mildly toxic by swallowing gross amounts of pure chemical.

Synonyms: CAS: 109-19-3 ✲ ISOVALERIC ACID, BUTYL ESTER ✲ n-BUTYL ISOPENTANOATE ✲ BUTYL-3-METHYLBUTYRATE

6-tert-BUTYL-m-CRESOL

Products: In germicides, perfumes, lubricating oils, and additives.

Use: As a disinfectant and a fixative.

Health Effects: Possible allergen. Irritant to skin. Moderate fire risk.

Synonyms: ✲ MBMC ✲ 6-tert-BUTYL-3-METHYLPHENOL

tert-BUTYL HYDROPEROXIDE

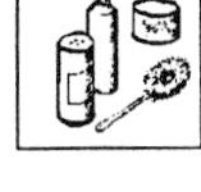

Products: In bleaching products and deodorizing agents.

Use: As cleanser or in cleaning products.

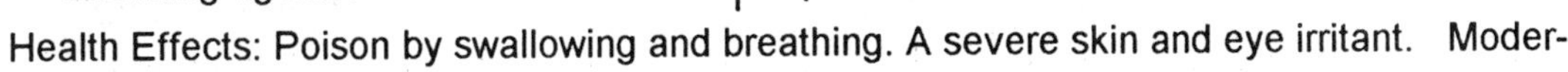

Health Effects: Poison by swallowing and breathing. A severe skin and eye irritant. Moderate fire risk.

Synonyms: CAS: 75-91-2 ✲ CADOX TBH ✲ PERBUTYL H ✲ 1,1-DIMETHYLETHYL HYDROPEROXIDE

n-BUTYL LACTATE

Products: In varnishes, inks, stencil pastes, perfumes, dry cleaning fluids, and adhesives.

Use: Various.

Health Effects: Possible allergen. A skin irritant. Toxic when breathed.

Synonyms: CAS: 138-22-7 ✲ 2-HYDROXYPROPANIC ESTER ✲ LACTIC ACID BUTYL ESTER

BUTYL MERCAPTAN

Products: Used in natural gas by power

Use: Warning odorant. Power

companies.

companies use one drop for every 1000 cubic feet. The normal nose can smell less than one per cent. At seven per cent the gas is explosive.

Health Effects: Has a strong offensive odor. An eye irritant.

Synonyms: CAS: 109-79-5 ✲ BUTANETHIOL

BUTYL OLEATE

Products: In water-proofing agents, compounds, coatings, polishes, and for outdoor equipment.

Use: As solvent or lubricant.

Health Effects: A possible allergen.

Synonyms: CAS: 142-77-8 ✲ WATER GUARD ✲ FAB GUARD

BUTYL PARABEN

Products: A wide variety of cosmetic and food products.

Use: An antifungal preservative; inhibits mold growth.

Health Effects: A skin irritant and possible allergen.

Synonyms: CAS: 94-26-8 ✲ BUTOBEN ✲ BUTYL CHEMOSEPT ✲ BUTYL PARASEPT ✲ BUTYL TEGOSEPT ✲ PARASEPT ✲ BUTYL-p-HYDROXYBENZOATE

p-tert-BUTYLPHENOL

Products: In lubricants, insecticides, motor oil additive, and petroleum oils and synthetic oils.

Use: Odorant.

Health Effects: An irritant to eyes and skin.

Synonyms: CAS: 98-54-4

n-BUTYL PROPIONATE

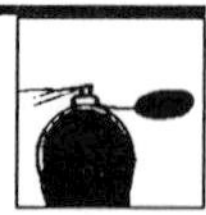

Products: A lacquer thinner additive and perfume ingredient.

Use: As an odorant and flavoring.

Health Effects: Skin and eye irritant. Flammable, moderate fire risk. Combustible.

Synonyms: CAS: 590-01-2

BUTYRIC ACID

Products: In perfume ingredients, disinfectants, gasoline odorant, butter, candy, caramels, and fruit seasonings.

Use: As a flavoring and odorant.

Health Effects: It is moderately toxic by swallowing large amounts of pure substance. Strong irritant to skin and tissue. Possible allergen. Corrosive material.

Synonyms: CAS: 107-92-6 ✲ n-BUTYRIC ACID ✲ BUTANOIC ACID ✲ ETHYLACETIC ACID ✲ PROPYLFORMIC ACID

BUTYROLACTONE

Products: Paint remover.

Use: Solvent

Health Effects: Toxic by swallowing.

Synonyms: CAS: 96-48-0 ✲ γ-BUTRYOLACTONE

"Everything in our world is composed of chemicals - animals, plants, and minerals, as well as the water we drink and the air we breathe."

American Chemical Society

CACODYLIC ACID

Products: In herbicide, soil sterilant, and for timber thinning.

Use: As a defoliant.

Health Effects: Toxic by swallowing. A skin and eye irritant. Possible carcinogen (may cause cancer).

Synonyms: CAS: 75-60-5 ❊ AGENT BLUE ❊ BOLLSEYE ❊ CHEXMATE ❊ DIMETHYL ARSENIC ACID ❊ ERASE ❊ SALVO

CADMIUM

Products: In pigments, enamels, photography, glazes, and batteries.

Use: Pigment, plating element, current carrier.

Health Effects: Flammable. A carcinogen (may cause cancer). Cadmium plating of food and beverage containers resulted in a number of outbreaks of gastroenteritis (food poisoning).

Synonyms: CAS: 7440-43-9 ❊ COLLOIDAL CADMIUM ❊ METALLIC ELEMENT OF ATOMIC NUMBER 48

CADMIUM TUNGSTATE

Products: In fluorescent paint.

Use: Produces crackle or crystal effects.

Health Effects: Toxic by breathing.

Synonyms: ❊ DAYGLOW

CAFFEINE

Products: In soft drinks (cola, orange), beverages (tea, coffee), and cocoa.

Use: As a stimulant.

Health Effects: GRAS (generally regarded as safe) by FDA when used at moderate levels. In excessive amounts it has been proven to be a human and animal poison by swallowing pure substance. Effects on the human body by swallowing include: ataxia (uncoordination), blood pressure elevation, convulsions, diarrhea, distorted perceptions, hallucinations, muscle contraction or spasticity, somnolence (general depressed activity), nausea or vomiting, tremors. A human teratogen (causing developmental abnormalities of the craniofacial and musculoskeletal systems), miscarriage and stillbirth. Human mutation (changes inherited characteristics) information reported. A possible carcinogen that caused cancer in animals. Large doses (above 1.0 gram) cause palpitation, excitement, insomnia, dizziness, headache, and vomiting. Too frequent excessive use of caffeine in tea or coffee may lead to digestive disturbances, constipation, palpitations, shortness of breath, and depressed mental states. It is also implicated in cardiac disorders and fibrocystic breast problems. One cup of drip-brewed coffee has 115 mg. One cup of tea has 40 mg. A twelve ounce cola drink is 46 mg. More than 300 mg. a day can overstimulate the central nervous system.

Synonyms: CAS: 58-08-2 ✲ CAFFEIN ✲ COFFEINE ✲ ELDIATRIC C ✲ GUARANINE ✲ NO-DOZ ✲ ORGANEX ✲ THEIN

CALCIUM ACETATE

Products: In baked goods, cake mixes, fillings, gelatins, packaging materials, puddings, sweet sauces, syrups, toppings, and sausage casings.

Use: As a processing aid, sequestrant (affects the final products appearance, flavor or texture), stabilizer, texturizing agent, thickening agent, and antimold agent.

Health Effects: (GRAS) Generally regarded as safe when used within limits specified by FDA.

Synonyms: CAS: 62-54-4 ✲ ACETATE of LIME ✲ BROWN ACETATE ✲ CALCIUM DIACETATE ✲ GRAY ACETATE ✲ LIME ACETATE ✲ LIME PYROLIGNITE ✲ SORBO-CALCIAN ✲ DIACETATE ✲ VINEGAR SALTS

CALCIUM ALGINATE

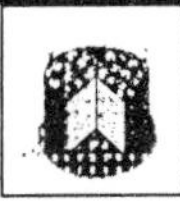

Products: In alcoholic beverages, baked goods, ice cream, confections, egg products, fats, frostings, canned fruits, gelatins, gravies, jams, oils, puddings, and sauces.

Use: As an emulsifier, stabilizer, and thickening agent.

Health Effects: GRAS (generally regarded as safe).

Synonyms: CAS: 9005-35-0 ✲ ALGIN ✲ ALGINIC ACID

CALCIUM ASCORBATE

Products: An antioxidant (slows reaction with oxygen) and preservative.

Use: Various.

Health Effects: GRAS (generally regarded as safe).

Synonyms: CAS: 5743-27-1

CALCIUM BENZOATE

Products: In margarine and oleomargarine.

Use: As a preservative.

Health Effects: GRAS (generally regarded as safe).

Synonyms: None found.

CALCIUM BROMATE

Products: In baked goods.

Use: As a dough conditioner and maturing agent.

Health Effects: GRAS (generally regarded as safe).

Synonyms: None found.

CALCIUM CARBONATE

Products: In baking powder, chewing gum, desserts (dry mix), dough, packaging materials, and wine. In steel, cement, plastic and over 800 industrial products. Mines in Mexico are a major source of the mineral.

Use: As an alkali, dietary supplement, dough conditioner, firming agent, modifier for chewing gum, nutrient, release agent for chewing gum, stabilizer, texturizing agent, chewing gum texturizer, and yeast food.

Health Effects: A severe eye and moderate skin irritant. GRAS (generally regarded as safe). It is considered the most concentrated and cheapest form of commercial calcium supplement. It is best absorbed when taken with food. The recommended daily allowance is about 800 mg.

Synonyms: CAS: 1317-65-3 ❊ AGRICULTURAL LIMESTONE ❊ AGSTONE ❊ ARAGONITE ❊ ATOMIT ❊ CALCITE ❊ CHALK ❊ DOLOMITE ❊ LIMESTONE ❊ LITHOGRAPHIC STONE ❊ MARBLE ❊ PORTLAND STONE

CALCIUM CHLORIDE

Products: In apple slices, baked goods, beverage bases (nonalcoholic), beverages (nonalcoholic), cheese, coffee, condiments, dairy product analogs, fruits (processed), fruit juices, gravies, jams (commercial), jellies (commercial), meat (raw cuts), meat products, milk (evaporated), pickles, plant protein products, potatoes (canned), poultry (raw cuts), relishes, sauces, tea, tomatoes (canned), and

Use: As as anticaking agent, antimicrobial agent, curing agent, firming agent, flavor enhancer, humectant (moisturizer), nutrient supplement, pickling agent, sequestrant (affects the final product appearance flavor or texture), stabilizer, texturizing agent, and thickening agent.

vegetable juices (processed).

Health Effects: In large amounts concentrated chemical is moderately toxic by swallowing. Possible carcinogen (may cause cancer) that caused tumors in animal studies. FDA states GRAS (generally regarded as safe) when used within limits.

Synonyms: CAS: 10043-52-4 ✲ CALPLUS ✲ CALTAC ✲ DOWFLAKE ✲ LIQUIDOW ✲ PELADOW ✲ SNOMELT ✲ SUPERFLAKE ANHYDROUS.

CALCIUM CITRATE

Products: In beans (lima), flour, and peppers.

Use: As a buffer (regulates the acidity or alkalinity), dietary supplement, firming agent, and nutrient.

Health Effects: GRAS (generally regarded as safe).

Synonyms: CAS: 813-94-5 ✲ LIME CITRATE ✲ TRICALCIUM CITRATE

CALCIUM DISODIUM EDTA

Products: In beverages (distilled alcoholic), beverages (fermented malt), cabbage (pickled), clams (cooked canned), corn (canned), crabmeat (cooked canned), cucumbers (pickled), dressings (nonstandardized), egg product (that is hard-cooked and consists, in a cylindrical shape, of egg white with an inner core of egg yoke), French dressing, lima beans (dried, cooked canned), margarine, mayonnaise, mushrooms, oleomargarine, pecan pie filling, pinto beans (processed dry), potato salad, potatoes (canned white), salad dressings, sauces, shrimp (cooked canned), soft drinks (canned carbonated), spreads (sandwich), and spreads (artificially colored and lemon-flavored or orange-flavored).

Use: As a preservative and sequestrant (binds constituents that affect the final product appearance, flavor or texture).

Health Effects: FDA states GRAS (generally regarded as safe) when used within stated limits.

Synonyms: ❖ CALCIUM DISODIUM EDETATE ❖ CALCIUM DISODIUM ETHYLENEDIAMINETETRAACETATE ❖ CALCIUM DISODIUM (ETHYLENEDINITRILO)TETRAACETATE

CALCIUM GLUCONATE

Products: In baked goods, dairy product analogs (simulations), gelatins, gels, puddings, and sugar substitutes.

Use: As a firming agent, formulation aid, sequestrant (binds constituents that affect the final product appearance, flavor or texture), stabilizer (used to keep a uniform consistency), texturizing agent, and thickening agent.

Health Effects: Possible allergen. GRAS (generally regarded as safe) when used within FDA limitations.

Synonyms: CAS: 299-28-5

CALCIUM HYDROXIDE

Products: In brick mortar, plasters, cements, depilatory (hair remover), disinfectants, water softening, and purification of sugar juices.

Use: As food additive buffer, firming agent, neutralizing agent, and miscellaneous general-purpose food chemicals.

Health Effects: It is mildly toxic by swallowing large amounts of pure substance. A severe eye irritant. A skin, nose, throat and respiratory system irritant. Causes dermatitis (skin rash or irritation). GRAS (generally regarded as safe) when used at moderate levels to accomplish the intended results.

Synonyms: CAS: 1305-62-0 ❖ BELL MINE ❖ CALCIUM HYDRATE ❖ HYDRATED LIME ❖ KEMIKAL ❖ LIME WATER ❖ SLAKED LIME

CALCIUM HYPOCHLORITE

Products: An algicide and bactericide for swimming pools. A deodorant, water purifier, fungicide, and bleaching agent.

Use: As a disinfectant and whitener.

Health Effects: Moderately toxic by swallowing. Can cause severe irritation of nose, throat and skin. Can cause fumes capable of producing pulmonary edema (lung damage). A mutagen (changes inherited characteristics). Dangerous fire risk.

Synonyms: CAS: 7778-54-3 ❖ CALCIUM OXYCHLORIDE ❖ BLEACHING POWDER ❖ CALCIUM HYPOCHLORIDE ❖ HTH ❖ HY-CHLOR ❖ LIME CHLORIDE ❖ PERCHLORON

CALCIUM IODATE

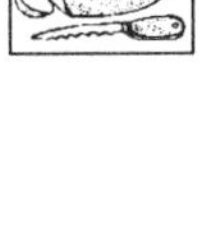

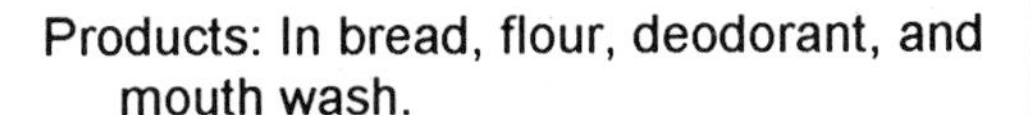

Products: In bread, flour, deodorant, and mouth wash.

Use: As a dough conditioner, maturing agent, food additive, and deodorizer.

Health Effects: GRAS (generally regarded as safe) by FDA when used within limits.

Synonyms: CAS: 7789-80-2

CALCIUM LACTATE

Products: In bread, cake (angel food), fruits (canned), meat food sticks, meringues, milk (dry powder), sausage, sausage (imitation), vegetables (canned), and whipped toppings.

Use: As a buffer, dough conditioner, firming agent, flavor enhancer, flavoring agent, leavening agent, nutrient supplement, stabilizer, thickening agent, additive, and yeast food.

Health Effects: GRAS (generally regarded as safe) by FDA when used within limits stated except for infant foods and formulas.

Synonyms: CAS: 814-80-2

CALCIUM LACTOBIONATE

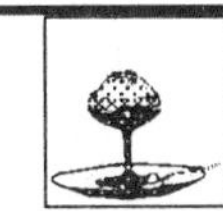

Products: In pudding mixes (dry).

Use: As a firming agent.

Health Effects: Harmless when used for intended purposes.

Synonyms: CAS: 5001-51-4 ✲ CALCIUM 4-(β-*d*-GALACTOSIDO)-*d*-GLUCONATE

CALCIUM NITRATE

Products: In fireworks, explosives, matches, and fertilizers.

Use: In pyrotechnics.

Health Effects: An irritant. Forms powerfully explosive mixtures with aluminum + ammonium nitrate + formamide + water, ammonium nitrate + hydrocarbon oils, ammonium nitrate + water soluble fuels and organic materials.

Synonyms: CAS: 10124-37-5

CALCIUM OXIDE

Products: In poultry feeds, sugar, insecticides, and fungicides.

Use: Sugar refining agent, food additive, dietary supplement, dough conditioner, nutrient, and yeast food. Before electricity an oxyhydrogen flame impinged on a cylinder of lime which caused a brilliant white light which was concentrated to a beam by a lens. The light was used as a spotlight for stage shows. Thus the phrase "in the limelight".

Health Effects: Very irritating to skin, eyes, nose and throat.

Synonyms: CAS: 1305-78-8 ❖ BURNT LIME ❖ CALCIA ❖ CALX ❖ LIME ❖ QUICKLIME

CALCIUM PEROXIDE

Products: In bakery products, seed disinfectants, tooth powders and pastes.

Use: As a dough conditioner, oxidizing agent, and antiseptic.

Health Effects: Irritating in concentrated form. A strong alkali. An oxidizer. GRAS (generally regarded as safe) when used in moderate amounts.

Synonyms: CAS: 1305-79-9 ❖ CALCIUM DIOXIDE ❖ CALCIUM SUPEROXIDE

CALCIUM PHOSPHATE, TRIBASIC

Products: In cereals, desserts, fats (rendered animal), flour, lard, packaging materials, table salt, and vinegar (dry). Ceramics, polishing agents, fertilizers, dentifrices, and meat tenderizer.

Use: As an anticaking agent, buffer, dietary supplement, fat rendering aid, nutrient, and stabilizer.

Health Effects: Skin and eye irritant. GRAS (generally regarded as safe) when used in moderate amounts.

Synonyms: CAS: 12167-74-7 ❖ PERCIPITATED CALCIUM PHOSPHATE ❖ TRICALCIUM PHOSPHATE ❖ CALCIUM ORTHOPHOSPHATE ❖ TERTIARY CALCIUM PHOSPHATE

CALCIUM PHOSPHIDE

Products: In rat poisons, fireworks, torpedoes, and signal fires.

Use: As a rodenticide and for pyrotechnics.

Health Effects: Dangerous fire risk. Highly toxic.

Synonyms: ❖ PHOTOPHOR

CALCIUM PHYTATE

Products: Sequestering agent (affects the final appearance, flavor or texture of product) to remove metals from wine and vinegar. In pharmaceuticals and

Use: As a binding agent, calcium supplement, and diet supplement.

nutrients.

Health Effects: Harmless when used for intended purpose.

Synonyms: ✲ HEXACALCIUM PHYTATE

CALCIUM PROPIONATE

Products: In cheese, confections, dough (fresh pie), fillings, and frostings, gelatins, jams, jellies, pizza crust, puddings, tobacco, and pharmaceuticals.

Use: As an antimicrobial agent, mold inhibitor, preservative, additive, and antifungal.

Health Effects: GRAS (generally regarded as safe).

Synonyms: CAS: 4075-81-4

CALCIUM PYROPHOSPHATE

Products: As a buffer, dietary supplement, neutralizing agent, and nutrient.

Use: For polishing agents in tooth paste or powder.

Health Effects: GRAS (generally regarded as safe).

Synonyms: CAS: 7790-76-3

CALCIUM RESINATE

Products: In waterproofers, paint driers, perfumes, cosmetics, enamels, and soaps.

Use: As a coating for fabrics, gel thickener, and detergents.

Health Effects: Flammable. Dangerous fire risk.

Synonyms: CAS: 9007-13-0 ✲ LIMED ROSIN

CALCIUM SORBATE

Products: In syrups (chocolate and fruit), fresh fruit salad, beverages, bakery goods, cheesecake, cheese, jellies, and salads (slaw, gelatin, macaroni, potato).

Use: As a mold retardant and preservative.

Health Effects: FDA states GRAS (generally regarded as safe) when used at moderate levels to accomplish the desired results.

Synonyms: None found.

CALCIUM STEARATE

Products: In beet sugar, candy (pressed), garlic salt, meat tenderizer, molasses (dry), salad dressing mix, vanilla, yeast, hair products, and paints.

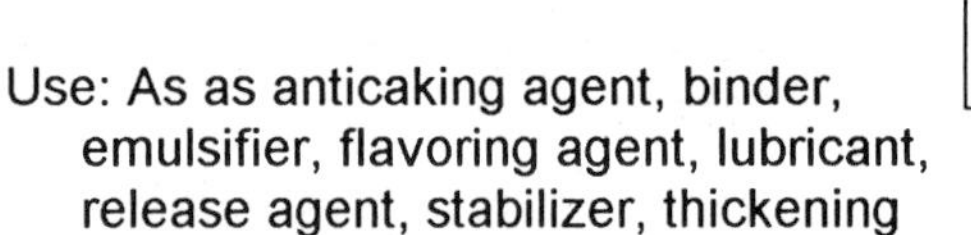

Use: As as anticaking agent, binder, emulsifier, flavoring agent, lubricant, release agent, stabilizer, thickening agent, and coloring agent.

Health Effects: FDA approves use at moderate levels to accomplish the desired results.

Synonyms: CAS: 1592-23-0

CALCIUM STEAROYL LACTATE

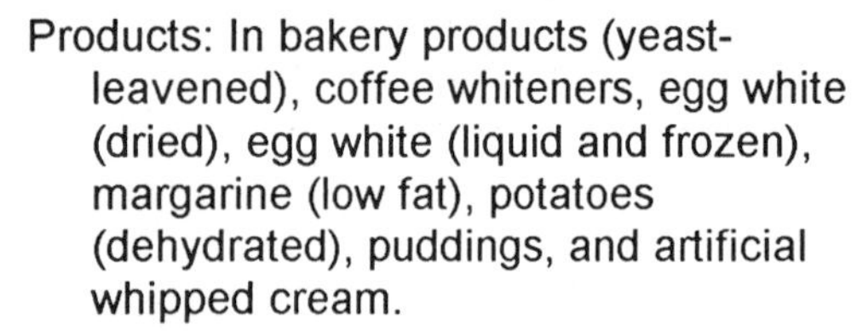

Products: In bakery products (yeast-leavened), coffee whiteners, egg white (dried), egg white (liquid and frozen), margarine (low fat), potatoes (dehydrated), puddings, and artificial whipped cream.

Use: As a dough conditioner, stabilizer, and whipping agent.

Health Effects: FDA approves limited use. Must conform to limitations.

Synonyms: ❖ CALCIUM STEAROYL-2-LACTATE

CALCIUM SULFATE

Products: In baked goods, canned potatoes, canned tomatoes, carrots (canned), confections, frostings, frozen dairy dessert mixes, frozen dairy desserts, gelatins, grain products, ice cream (soft serve), lima beans (canned), pasta, peppers (canned), puddings, wine (sherry), paints (pigment, filler), surgical casts, gypsum board (drywall), and quick-setting cements.

Use: As an anticaking agent, color, coloring agent, dietary supplement, dough conditioner, dough strengthener, drying agent, firming agent, flour treating agent, leavening agent, nutrient supplement, sequestrant, stabilizer, texturizing agent, thickening agent, and yeast food.

Health Effects: Harmless when used for intended purposes.

Synonyms: CAS: 7778-18-9/10101-41-4 ❖ GYPSUM ❖ PLASTER of PARIS

CALCIUM SULFIDE

Products: In luminous paint, depilatories, acne medication, oil additive.

Use: As a preservative in paints and in skin products.

Health Effects: Poison via breathing. Possible allergen. Irritating to skin, nose and throat.

Synonyms: CAS: 20548-54-3 ❖ CALCIC LIVER OF sulfur ❖ HEPAR CALCIS ❖ OLDHAMITE ❖ CALCIUM SULPHIDE

CALCIUM SULFITE

Products: As a sugar disinfectant; brewing disinfectant.

Use: For cleansing and preserving. It prevents discoloring.

Health Effects: Harmless when used for intended purpose.

Synonyms: None found.

CALCIUM THIOGLYCOLLATE

Products: In depilatory and hair waving products.

Use: For hair removal and hair permanents.

Health Effects: Possible allergen. Possible skin irritant.

Synonyms: None found.

CAMPHENE

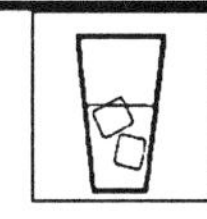

Products: In beverages, ice desserts, candies, and bakery goods.

Use: Adds a nutmeg, spice-like fragrance.

Health Effects: GRAS (generally regarded as safe) when used in moderate amounts.

Synonyms: CAS: 79-92-5

CAMPHOR

Products: In skin lotion, moth and mildew proofings, toothpowders, lacquers, and insecticides.

Use: For fragrances, flavorings, and conditioners.

Health Effects: Toxic by breathing, swallowing, and absorption through skin. Skin irritant. Swallowing causes nausea, vomiting, dizziness, and convulsions.

Synonyms: CAS: 76-22-2 ❖ GUM CAMPHOR ❖ 2-CAMPHANONE

CANANGA OIL

Products: In carbonated drinks (cola, ginger ale), ice desserts, candies, and bakery goods.

Use: For spice and food flavorings.

Health Effects: Possible allergen. GRAS (generally regarded as safe) when used in moderate amounts.

Synonyms: None found.

CANDELILLA WAX

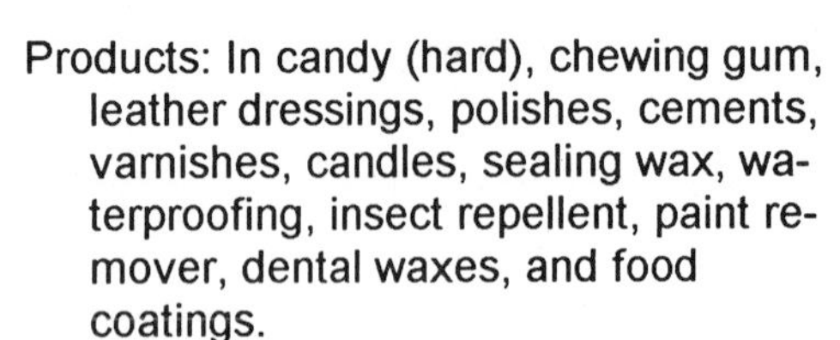

Products: In candy (hard), chewing gum, leather dressings, polishes, cements, varnishes, candles, sealing wax, waterproofing, insect repellent, paint remover, dental waxes, and food coatings.

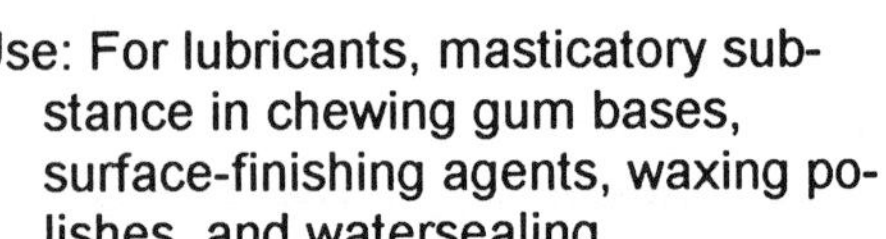

Use: For lubricants, masticatory substance in chewing gum bases, surface-finishing agents, waxing polishes, and watersealing.

Health Effects: FDA states GRAS (generally regarded as safe) when used in moderate amounts.

Synonyms: CAS: 8006-44-8

CANTHAXANTHIN

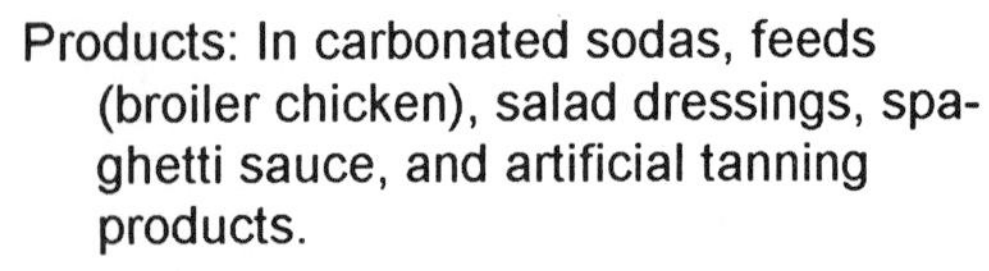

Products: In carbonated sodas, feeds (broiler chicken), salad dressings, spaghetti sauce, and artificial tanning products.

Use: As color additive (reddish) for food and drugs.

Health Effects: Swallowing pure substance may cause night blindness. FDA permits limited use in foods.

Synonyms: CAS: 514-78-3 ❊ CANTHA ❊ β-CAROTENE-4,4′-DIONE ❊ 4,4′-DIKETO-β-CAROTENE

CAPRIC ACID

Products: In perfumes, lipstick, butter, fruits, coconut, liquor, cheese flavorings, beverages, ice and gelatin desserts, bakery goods, chewing gum, and shortenings.

Use: For flavorings and fragrances.

Health Effects: Harmless when used for intended purposes.

Synonyms: CAS: 334-48-5 ❊ DECANOIC ACID ❊ DECYLIC ACID

CAPROIC ACID

Products: In varnish products and in food products.

Use: For resins and flavorings.

Health Effects: Moderately toxic by swallowing pure acid. Possible mutagen (changes inherited characteristics). Corrosive. A skin and eye irritant. GRAS (generally regarded as safe) for limited and intended uses.

Synonyms: CAS: 142-62-1 ✲ HEXANOIC ACID ✲ HEXYLIC ACID ✲ HEXOIC ACID

CAPSAICIN

Products: In spicy seasonings. In self-defense sprays.

Use: Flavor found in chili peppers.

Health Effects: Irritating and burning sensation to nose and throat that cannot by alleviated by water. However, milk, sour cream or yogurt has cooling effect.

Synonyms: ✲ HOT PEPPERS ✲ TABASCO

CAPTAN

Products: In almonds, animal feed, apples, beans, beef, beets, broccoli, cabbage, carrots, corn, garlic, kale, lettuce, peaches, peas, pork, potatoes, raisins, spinach, and strawberries. In paints, plastics, leather, and fabrics.

Use: As fungicide, preservative, and bacteriostat.

Health Effects: Pure substance is moderately toxic to humans by swallowing large amounts. Human mutagen (changes inherited characteristics). Produce should be washed well before consuming!

Synonyms: CAS: 133-06-2 ✲ AGROSOL S ✲ AMERCIDE ✲ BANGTON ✲ CAPTANE ✲ CAPTAN-STREPTOMYCIN 7.5-0.1 POTATO SEED PIECE PROTECTANT ✲ CAPTEX ✲ FLIT 406 ✲ HEXACAP ✲ KAPTAN ✲ MERPAN ✲ ORTHOCIDE ✲ OSOCIDE ✲ VANICIDE

CARAMEL

Products: In baked goods, colas, root beer, butterscotch, chocolate, fruits, rums, maple, walnut, vanilla syrups, ice desserts, bakery goods, and meats. In cosmetics.

Use: As color and flavoring additive.

Health Effects: A possible allergen. FDA states GRAS (generally regarded as safe) when used at moderate levels.

Synonyms: CAS: 8028-89-5 ✲ CARAMEL COLOR

CARAWAY OIL

Products: In bakery products, beverages (nonalcoholic), condiments, ice cream products, grape licorice, anisette, rye flavorings, liquors, and soap perfume.

Use: As a flavoring and perfuming agent.

Health Effects: In large amounts pure oil is moderately toxic by swallowing and skin contact. A skin irritant. GRAS (generally regarded as safe) when used at moderate levels.

Synonyms: CAS: 8000-42-8 ✲ KUEMMEL OIL (GERMAN) ✲ OIL of CARAWAY

CARBANOLATE

Products: On animal feed, bananas, beans, citrus fruit, coffee, hops (dried), peanuts, pecans, potatoes, sorghum, soybeans, sugar beets, sugarcane, and sweet potatoes.

Use: As insecticide and nematocide (kills parasitic worms).

Health Effects: Deadly poison by swallowing or skin contact. Human mutagen (changes inherited characteristics). In 1985 over 150 people in California exhibited toxic effects from eating watermelons contaminated with aldicarb. FDA permits limited use. On EPA Extremely Hazardous Substances list.

Synonyms: CAS: 116-06-3 ✲ ALDECARB ✲ ALDICARB ✲ AMBUSH ✲ TEMIC ✲ TEMIK

CARBON BLACK

Products: In printing inks, carbon paper, typewriter ribbons, and paint pigments.

Use: As color additive.

Health Effects: Mildly toxic by swallowing, breathing, and skin contact. Possible carcinogen (may cause cancer). A nuisance dust in high concentrations. While it is true that the tiny particulates of carbon black contain some molecules of carcinogenic materials, the carcinogens are apparently held tightly and are not released by hot or cold water, gastric juices, or blood plasma. FDA no longer permits use in food.

Synonyms: CAS: 1333-86-4 ✲ ACETYLENE BLACK ✲ CHANNEL BLACK ✲ FURNACE BLACK ✲ LAMP BLACK

CARBON DIOXIDE

Products: In beverages (carbonated), fruit, meat, poultry, and wine.

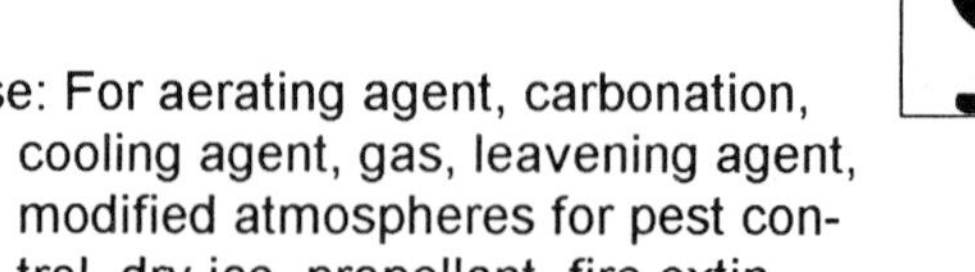

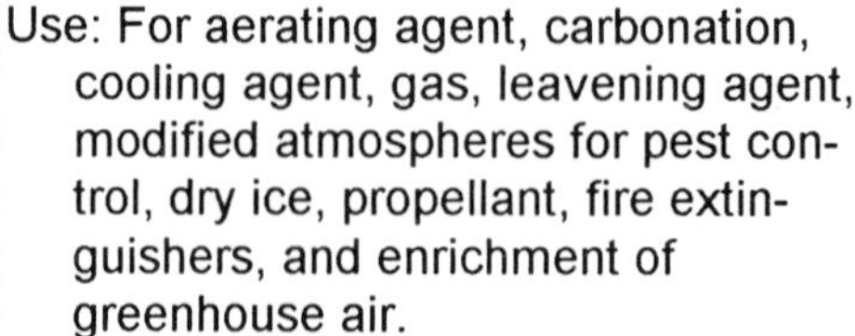

Use: For aerating agent, carbonation, cooling agent, gas, leavening agent, modified atmospheres for pest control, dry ice, propellant, fire extinguishers, and enrichment of greenhouse air.

Health Effects: An asphyxiant. Contact of carbon dioxide snow with the skin can cause burns. FDA states GRAS (generally regarded as safe) when used at moderate levels to accomplish the desired results.

Synonyms: CAS: 124-38-9 ✲ CARBONIC ACID GAS ✲ CARBONIC ANHYDRIDE

CARBON TETRACHLORIDE

Products: In spot remover, fire extinguishers, dry cleaning products, fumigants, solvents, and pesticides.

Use: Various.

Health Effects: A poison by swallowing. Mildly toxic by breathing. An eye and skin irritant. Damages liver, kidneys, and lungs. It has a narcotic action resembling chloroform. Probably a carcinogen (causes cancer). Banned from all uses in 1996 due to ozone depletion concerns.

Synonyms: CAS: 56-23-5 ✲ BENZINOFORM ✲ CARBON CHLORIDE ✲ CARBON TET ✲ METHANE ✲ NECATORINE ✲ TETRAFORM ✲ TETRASOL ✲ METHANE TETRACHLORIDE

CARBOXYMETHYLCELLULOSE

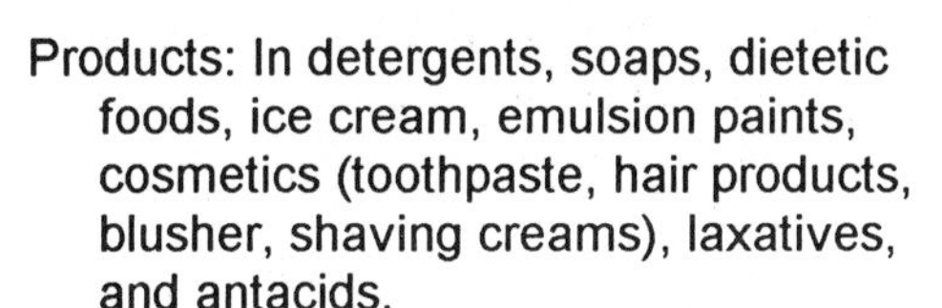

Products: In detergents, soaps, dietetic foods, ice cream, emulsion paints, cosmetics (toothpaste, hair products, blusher, shaving creams), laxatives, and antacids.

Use: As emulsifier, stabilizer, and foaming agent.

Health Effects: In excessive amounts it is mildly toxic by swallowing pure substance. It caused animal reproductive effects (infertility, or sterility, or birth defects). Possible carcinogen (may cause cancer). It migrates to food from packaging materials.

Synonyms: CAS: 9004-32-4 ✲ CARMETHOSE ✲ CELLUGEL ✲ AQUAPLAST ✲ CELLPRO

CARMINE

Products: In dyes, inks, and food color.

Use: As red color additive.

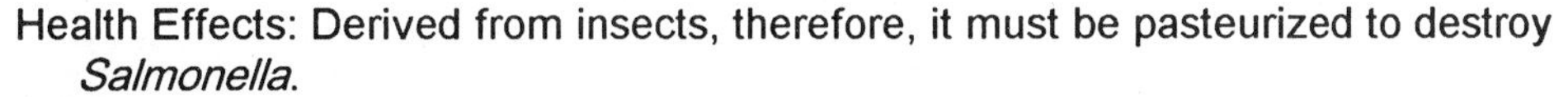

Health Effects: Derived from insects, therefore, it must be pasteurized to destroy *Salmonella*.

Synonyms: CAS: 1390-65-4 ✲ B ROSE LIQUID ✲ CARMINIC ACID

CARNAUBA WAX

Products: In baked goods, baking mixes, candy (soft), chewing gum, confections, frostings, fruit juices, fruit juices (processed), fruits (fresh), fruits (processed), gravies, sauces, texturizer, and depilatory (hair remover).

Use: An anticaking agent, candy glaze, candy polish, formulation aid, lubricant, release agent, makeup base, blusher, mascara, lipstick, hair remover, antiperspirant, and liquid makeup.

Health Effects: FDA states GRAS (generally regarded as safe) when used at moderate levels to accomplish the desired results.

Synonyms: CAS: 8015-86-9 ❊ BRAZIL WAX

CARRAGEEN

Products: Derived from dried weed of sea weed. In dairy products, dessert gels (water), jelly (low calorie), meat (restructured), poultry, cosmetic oil, chocolate, toothpaste, ice cream, and chocolate milk.

Use: As a binder, emulsifier, extender, gelling agent, stabilizer, and thickening agent.

Health Effects: GRAS (generally regarded as safe) by FDA when used within limits. Pure substance is a suspected carcinogen (may cause cancer) that caused tumors in animals.

Synonyms: CAS: 9000-07-1 ❊ AUBYGEL ❊ AUBYGUM ❊ CARASTAY ❊ CARRAGEENAN GUM ❊ GALOZONE ❊ GELCARIN ❊ GELOZONE ❊ GENU ❊ GENUGEL ❊ IRISH GUM ❊ IRISH MOSS EXTRACT ❊ IRISH MOSS GELOSE

CARVACROL

Products: In perfume, fungicide, disinfectants, citrus, fruit, mint, and spice flavorings for beverage, ice dessert, candies, bakery goods, and spicy condiments.

Use: As a flavoring and as a germicide.

Health Effects: Poison by swallowing large amounts of concentrated chemical. Moderately toxic by skin contact. A severe skin irritant. Combustible liquid. FDA approves use at moderate levels to accomplish the desired results.

Synonyms: CAS: 499-75-2 ❊ 2-p-CYMENOL ❊ ISOPROPYL-o-CRESOL ❊ 5-ISOPROPYL-2-METHYLPHENOL ❊ ISOTHYMOL ❊ 2-METHYL-5-ISOPROPYLPHENOL ❊ o-THYMOL

d-CARVONE

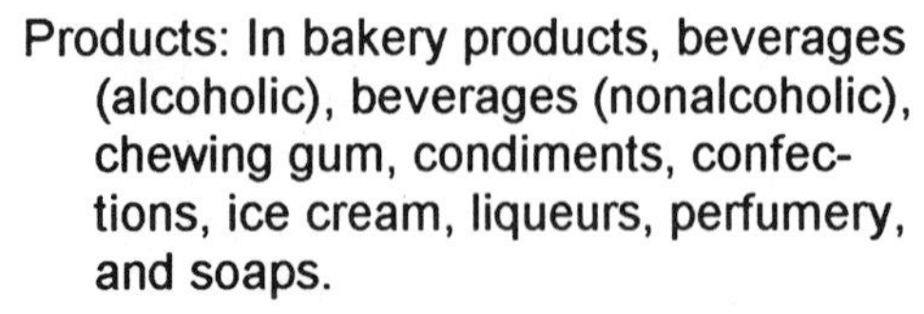

Products: In bakery products, beverages (alcoholic), beverages (nonalcoholic), chewing gum, condiments, confections, ice cream, liqueurs, perfumery, and soaps.

Use: As a flavoring and perfuming agent. This (*d*-) form is main constituent of caraway and dill oils.

Health Effects: In large amounts the pure chemical is poison by swallowing and by skin contact. A skin irritant. FDA approves use at moderate levels to accomplish the desired results.

Synonyms: CAS: 2244-16-8 ❊ (+)-CARVONE ❊ d(+)-CARVONE ❊ (S)-CARVONE ❊ (S)-(+)-CARVONE

l-CARVONE

Products: In beverages, ice desserts, candies, gum, bakery goods, perfumes, toilet soaps, OTC (over the counter) medications, candy mints, and breath mints.

Use: Flavoring for liqueur, mint, and spices. This (*l*-) form occurs principally in spearmint oil.

Health Effects: The concentrated chemical is moderately toxic by swallowing gross amounts. FDA approves use at moderate levels to accomplish the desired results.

Synonyms: CAS: 6485-40-1 ✲ (-)-CARVONE ✲ (R)-CARVONE

CASHEW GUM

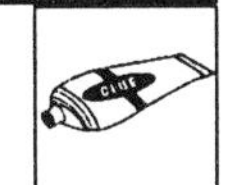

Products: In inks, insecticides, glues, tanning agent, varnishes, and bookbinders' gum.

Use: Binder derived from exudate from cashew nut tree.

Health Effects: Possible allergen. Harmless when used for intended purposes.

Synonyms: ✲ ANACARDIUM GUM

CASHEW NUTSHELL OIL

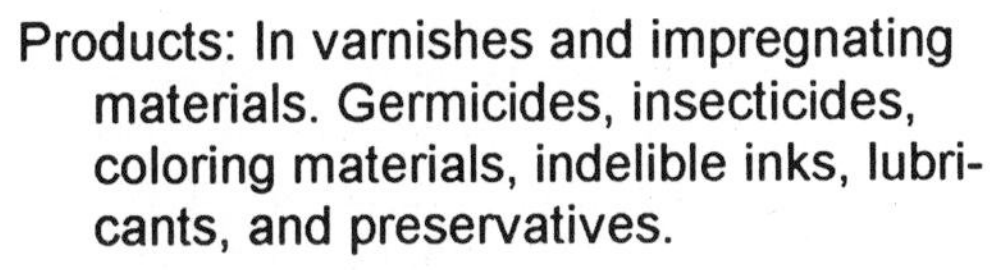

Products: In varnishes and impregnating materials. Germicides, insecticides, coloring materials, indelible inks, lubricants, and preservatives.

Use: Oil derived from liquid found between layers of nutshells.

Health Effects: Possible allergen. Strong irritant in natural state. Harmless when used for intended purposes.

Synonyms: ✲ CASHEW NUTSHELL LIQUID

CASEIN

Products: A nutritious milk protein containing all the essential amino acids. For cheesemaking, interior paints, adhesives for laminates (plywood), foods and feeds, and dietetic preparations.

Use: Thickener, whitening agent, binder, coatings, glues, and sizing.

Health Effects: Probably harmless when used for intended purposes.

Synonyms: CAS: 9005-46-3 ✲ SODIUM CASEINATE ✲ CASEIN AND CASEINATE ✲ NUTROSE

CASSIA OIL

Products: In perfumes, medications, and soaps.

Use: For flavorings, odorants, and laxatives.

Health Effects: Possible allergen. Poison by skin contact. Moderately toxic by swallowing. A human skin irritant. GRAS (generally regarded as safe) when used in moderate amounts.

Synonyms: CAS: 8007-80-5 ❊ ARTIFICIAL CINNAMON OIL ❊ CINNAMON BARK OIL ❊ CINNAMON BARK OIL, CEYLON TYPE (FCC) ❊ CINNAMON OIL

CASTOR OIL

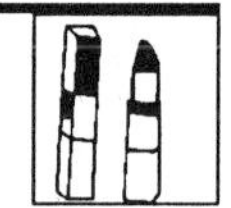

Products: In candy (hard), vitamin and mineral tablets, lipstick, laxatives, bath oils, soaps, hair products, solid perfumes, nail polish, and nail polish remover.

Use: As an antisticking agent. In coatings and medications.

Health Effects: Moderately toxic by swallowing large amounts of pure oil. An allergen. An eye irritant. Combustible when exposed to heat. Spontaneous heating may occur. Use at moderate levels to accomplish the desired results.

Synonyms: CAS: 8001-79-4 ❊ AROMATIC CASTOR OIL ❊ CASTOR OIL AROMATIC ❊ COSMETOL ❊ CRYSTAL O ❊ GOLD BOND ❊ OIL of PALMA CHRISTI ❊ PHORBYOL ❊ RICINUS OIL ❊ TANGANTANGAN OIL ❊ TURKEY RED OIL

CATIONIC IMIDAZOLINES

Products: In cosmetics and cleaners.

Use: As emulsifier (stabilizes and maintains mixes to aid in suspension of oily liquids), antistatic agent, and water displacer.

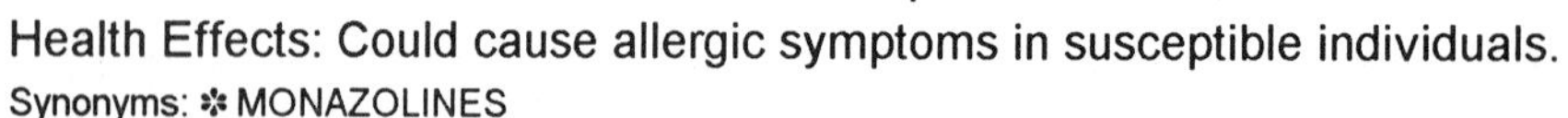

Health Effects: Could cause allergic symptoms in susceptible individuals.

Synonyms: ❊ MONAZOLINES

CELERY SEED OIL

Products: In meat, salads, sauces, soups, beverages, sausage, bakery goods, candies, condiments, syrups, fruit, nut, root beer, vanilla, gum, and pickles.

Use: For flavoring and seasoning in food and drink.

Health Effects: FDA states GRAS (generally regarded as safe) when used for intended purposes.

Synonyms: None found.

CETYL ALCOHOL

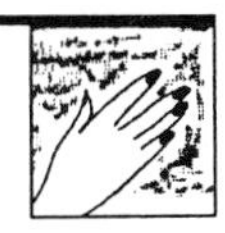

Products: In facial makeup, hair products, blushes, mascara, lipstick, nail polish remover, deodorant, and baby skin products.

Use: Solvent; aids in mixing ingredients..

Health Effects: Moderately toxic by swallowing. An eye and skin irritant. Possible allergen.

Synonyms: CAS: 36653-82-4 ❊ 1-HEXADECANOL ❊ ADOL ❊ ALCOHOL C-16 ❊ CETYLOL ❊ ETHAL ❊ ETHOL ❊ PALMITYL ALCOHOL

CETYL PYRIDINIUM CHLORIDE

Products: In cough lozenges and syrups.

Use: As an antibacterial.

Health Effects: Poison by swallowing gross amounts of concentrated chemical. Moderately toxic by skin contact. A skin and eye irritant.

Synonyms: CAS: 123-03-5 ❊ CEEPRYN ❊ CEPACOL ❊ CEPACOL CHLORIDE ❊ CETAMIUM ❊ DOBENDAN ❊ MEDILAVE ❊ PRISTACIN ❊ PYRISEPT

CHAMOMILE OIL

Products: In hair products, conditioners, rinses, and skin products.

Use: Oil distilled from flowers.

Health Effects: A mild allergen. A skin irritant. GRAS (generally regarded as safe) when used in moderate amounts.

Synonyms: CAS: 8002-66-2 ❊ CAMOMILE OIL GERMAN ❊ CHAMOMILE-GERMAN OIL ❊ GERMAN CHAMOMILE OIL ❊ HUNGARIAN CHAMOMILE OIL

CHAMOMILE OIL (ROMAN)

Products: In beverages, ice desserts, candies, bakery goods, fruit, chocolate, vermouth.

Use: Flavoring of vanilla, maple, berry, or spice.

Health Effects: A mild allergen. A skin irritant. Combustible when heated. GRAS (generally regarded as safe) when used at a level not in excess of the amount reasonably required to accomplish the desired results.

Synonyms: CAS: 8015-92-7 ❊ CAMOMILE OIL, ENGLISH TYPE

CHARCOAL, ACTIVATED

Products: For fats (refined animal), grape juice (black), grape juice (red), sherry

Use: As decolorizing agent, odor-removing agent, purification agent in

(cocktail), sherry (pale dry), and wine. fat and beverage processing, and taste-removing agent.

Health Effects: It can cause a dust irritation, particularly to the eyes, nose and throat. Combustible when exposed to heat. Dust is flammable and explosive when exposed to heat or flame. GRAS (generally regarded as safe) when used within limitations.

Synonyms: CAS: 64365-11-3 ❖ ACTIVATED CARBON ❖ CARBON, ACTIVATED ❖ AMORPHOUS CARBON

CHICLE

Products: Chewing gum.

Use: As a softener. Derived from latex of sapodilla tree in Mexico and Central America.

Health Effects: Swallowing should be avoided.

Synonyms: ❖ THERMOPLASTIC SUBSTANCE

CHLORAL HYDRATE

Products: In medications and liniments.

Use: As a sedative, anesthetic, and narcotic.

Health Effects: Overdose is toxic. Hypnotic drug. Damaging to eyes.

Synonyms: CAS: 302-17-0 ❖ KNOCKOUT DROPS ❖ TRICHLOR ACETALDEHYDE ❖ HYDRATED TRICHLOROETHYLIDENE GLYCOL

CHLORBENSIDE

Products: Acaricide.

Use: Kills mange mites.

Health Effects: Toxic by swallowing. A skin irritant.

Synonyms: CAS: 103-17-3 ❖ CHLOROCIDE ❖ CHLOROPARACIDE ❖ CHLOROBENZIDE ❖ CHLOROPARACIDE

CHLORDANE

Products: Insecticide.

Use: As pesticide and fumigant.

Health Effects: Suspected carcinogen that caused cancer in animals. Poison to humans by swallowing. Moderately toxic by skin contact. Effects by swallowing or skin contact are: tremors, convulsions, excitement, ataxia (loss of muscle coordination), and gastritis. A human mutagen (changes inherited characteristics). Combustible liquid. It is no longer permitted for use as a termiticide in homes.

Synonyms: CAS: 57-74-9 ❖ ASPON-CHLORDANE ❖ BELT ❖ CHLORDAN ❖ γ-CHLORDAN ❖ CHLORINDAN ❖ CHLOR KIL ❖ CHLORODANE ❖ CHLORTOX ❖ TOXICHLOR

CHLORINE

Products: For laundry bleach, bath cleansers, tile cleansers, and cleaning liquids.

Use: As an antimicrobial agent, bleaching agent, oxidizing agent.

Health Effects: Moderately toxic by breathing. A human mutagen (changes inherited characteristics). Some respiratory system effects by breathing are: changes in the trachea or bronchi, emphysema, chronic pulmonary edema (lung damage) or congestion. A strong irritant to eyes, nose, and throat.

Synonyms: CAS: 7782-50-5 ✲ BERTHOLITE ✲ CHLORINE MOL. ✲ MOLECULAR CHLORINE

α-CHLOROACETOPHENONE

Products: Spray and aerosol tear gas.

Use: As riot control agent; self-defense products.

Health Effects: Strong irritant to eyes and tissue as gas or liquid.

Synonyms: CAS: 532-27-4 ✲ MACE ✲ CHEMICAL MACE ✲ PHENACYCLCHLORIDE ✲ PHENYL CHLOROMETHYL KETONE

CHLOROFLUOROCARBON

Products: Spray containers and vehicle air conditioners

Use: Aerosol propellants and refrigerant.

Health Effects: High concentrations cause narcosis and anesthesia in humans. Harmful effects are eye irritation and liver changes. Poison by breathing. Exposure could cause blindness. It is believed to be depleting stratospheric (upper atmosphere) ozone layer, thus causing global warming. Ozone layer depletion has also resulted in an increase in skin cancers and cataracts. In the next few years all products containing CFCs will have been reformulated and CFC use will be prohibited.

Synonyms: CAS: 75-69-4 ✲ CFC ✲ TRICHLORO MONOFLUORO METHANE ✲ FREON 11 ✲ ARCTON 9 ✲ FRIGEN 11 ✲ HALOCARBON 11

CHLORO-IPC

Products: Herbicide.

Use: A pre-emergent sprayed on potatoes to prevent sprouting.

Health Effects: Hazard by swallowing. Possible carcinogen (may cause cancer). A human mutagen (changes inherited characteristics).

Synonyms: CAS: 101-21-3 ✲ BUD-NIP ✲ PREVENTOL ✲ TATERPEX ✲ CHLORPROPHAME

4-CHLORO-3-METHYLPHENOL

Products: In glues, gum, paints, inks, textiles, and leather goods.

Use: As a germicide and preservative.

Health Effects: Irritant to skin.

Synonyms: ✲ 4-CHLORO-1-HYDROXY-3-METHYLBENZENE ✲ p-CHLORO-m-CRESOL

CHLOROPHYLL

Products: In casings, fats (rendered), oleomargarine, shortening, soaps, toothpastes, cleaning products, chewing gum, waxes, liquors, cosmetics, perfumes, and dentifrices.

Use: As coloring agent and deodorant.

Health Effects: Safe, natural, green pigment present in all plants except fungi and bacteria.

Synonyms: CAS: 1406-65-1 ✲ BIOPHYLL ✲ DAROTOL ✲ DEODOPHYLL ✲ GREEN CHLOROPHYL

CHLOROPICRIN

Products: In fumigants, insecticides, rat exterminator and tear gas.

Use: As fungicide, pesticide, and rat poison.

Health Effects: Very toxic by swallowing and breathing. A strong irritant.

Synonyms: CAS: 76-06-2 ✲ NITROCHLOROFORM ✲ TRICHLORONITROMETHANE ✲ CHLORPICRIN

CIGARETTE TAR

Products: All cigarette brands.

Use: Smoking materials.

Health Effects: The tar contains various compounds, especially benzopyrene, found in coal tar which are carcinogens (causes cancer).

Synonyms: None found.

CINNAMALDEHYDE

Products: In bakery products, beverages (nonalcoholic) colas, fruit, liquors, rum, spices, chewing gum, condiments, confections, ice cream products, meat, mouthwashs, and toothpastes.

Use: In flavors and perfumery.

Health Effects: Moderately toxic by swallowing excessive amounts of concentrated chemical. A possible allergen. A severe human skin irritant. GRAS (generally regarded as safe) when used in moderate amounts.

Synonyms: CAS: 104-55-2 ✽ BENZYLIDENEACETALDEHYDE ✽ CASSIA ALDEHYDE ✽ CINNAMAL ✽ CINNAMYL ALDEHYDE ✽ CINNIMIC ALDEHYDE ✽ PHENYLACROLEIN

CINNAMIC ACID

Products: In beverages, ice cream, candies, bakery goods, chewing gum, and suntan products.

Use: In flavorings and perfumes. An ingredient to affect the taste or smell of final product.

Health Effects: In large amounts the pure substance is moderately toxic by swallowing. A possible allergen. A skin irritant. Harmless when used in moderate amounts.

Synonyms: CAS: 621-82-9 ✽ PHENYLACRYLIC ACID ✽ tert-β-PHENYLACRYLIC ACID ✽ 3-PHENYLACRYLIC ACID ✽ 3-PHENYLPROPENOIC ACID ✽ 3-PHENYL-2-PROPENOIC ACID

CINNAMON LEAF OIL

Products: In bakery products, beverages (nonalcoholic), chewing gum, condiments, ice cream, meat, and pickles.

Use: As a flavoring for perfume and toothpaste.

Health Effects: Possible allergen. FDA states GRAS (generally regarded as safe) when used in moderate amounts.

Synonyms: ✽ CINNAMON LEAF OIL, Ceylon ✽ CINNAMON LEAF OIL, Seychelles

CINNAMYL ALCOHOL

Products: In soaps and cosmetics.

Use: As a scenting agent in perfumery, particularly for lilac and other floral scents.

Health Effects: Moderately toxic by swallowing. A skin irritant.

Synonyms: CAS: 104-54-1 ✽ CINNAMIC ALDEHYDE ✽ CINNAMALDEHYDE ✽ 3-PHENYL PROPENAL ✽ 3-PHENYLALLYL ALCOHOL ✽ STYRONE ✽ STYRYL CARBINOL

CITRIC ACID

Products: In beef (cured), chili con carne, cured meat food product, fats (poultry), fruits (frozen), lard, meat (dried), pork (cured), pork (fresh), potato sticks, potatoes (instant), poultry, sausage (dry), sausage (fresh pork),

Use: As a preservative, flavoring agent, sequestrant (affects the appearance, flavor or texture of final product), neutralizes lye in vegetable peeling, curing of meats, prevents darkening of cut fruits and vegetables. A tart

shortening, wheat chips, and wine.

flavoring. A buffer that controls the acidity of jams, ice desserts and candies. Naturally abundant in citrus fruits and berries.

Health Effects: Mildly toxic by swallowing great amounts of concentrated acid. A severe eye and moderate skin irritant, some allergenic effects. GRAS (generally regarded as safe) when used within limitations. USDA states limitations of use.

Synonyms: CAS: 77-92-9 ✲ ACILETTEN ✲ CITRETTEN ✲ CITRO

CITRONELLAL

Products: In bakery products, beverages (nonalcoholic), chewing gum, confections, gelatin desserts, ice cream, meat, and puddings.

Use: As a flavoring agent.

Health Effects: Use at a moderate level to accomplish the desired results.

Synonyms: CAS: 106-23-0 ✲ 3,7-DIMETHYL-6-OCTENAL

CITRONELLA OIL

Products: Insect repellent, soap, perfumery, and disinfectants.

Use: As a mosquito repellent for candles and odorants.

Health Effects: May not be used on food crops.

Synonyms: CAS: 8000-29-1

CLARY OIL

Products: In beverages, vermouth wine, root beer, cream soda, butter, fruits, licorice, black cherry, grape, ice desserts, candies, and bakery goods.

Use: As a perfume fixative. A food and beverage spice.

Health Effects: GRAS (generally regarded as safe) when used in moderate amounts.

Synonyms: ✲ CLARY SAGE OIL ✲ OIL of MUSCATEL

CLOVE LEAF OIL MADAGASCAR

Products: In bakery products, beverages (alcoholic), beverages (nonalcoholic), root beer, cinnamon flavorings, cakes, chewing gum, condiments, confections, cookies, fruit punches, fruits

Use: As a food flavoring agent, dental analgesic (kills pain), and dental germicide.

(spiced), gelatin desserts, ice cream, marinades, meat, meat sauces, pickles, puddings, relishes, and sauces. Also in soaps and dental antiseptics and pharmaceutical products.

Health Effects: The pure substance in large amounts is moderately toxic by swallowing and skin contact. A severe skin irritant and possible allergen. FDA states GRAS (generally regarded as safe) when used in moderate amounts.

Synonyms: CAS: 8015-97-2 ✲ CLOVE LEAF OIL ✲ OILS, CLOVE LEAF

COAL TAR

Products: For roads, waterproofing, paints, pipe coatings, insulation, pesticides, sealants, adhesives, hair coloring products, and makeup.

Use: As surfacing, coatings, dyes, insecticides, personal grooming, hair, and cosmetic products.

Health Effects: A human carcinogen. Toxic by breathing. Frequently causes skin irritation and allergic reactions.

Synonyms: CAS: 8007-45-2 ✲ COAL TAR PITCH VOLATILES ✲ LAVATAR ✲ POLYTAR BATH ✲ SYNTAR

COCAINE

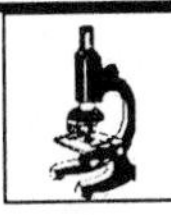

Products: Used in medicine as a local narcotic anesthetic.

Use: As anesthetic; possession is illegal in U.S.

Health Effects: A poison by swallowing and possibly other routes. Central nervous system effects by swallowing and possibly other routes are general anesthesia, hallucinations, distorted perceptions, possible cardiac failure and convulsions. An eye irritant. An abused controlled substance. Use leads to addiction and habit.

Synonyms: CAS: 50-36-2 ✲ METHYL BENZOYLECGONINE ✲ COKE ✲ CORINE ✲ ERITROXILINA ✲ GOLD DUST ✲ NEUROCAINE

COCOA BUTTER

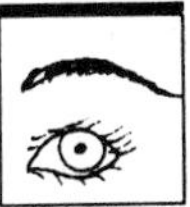

Products: In emollient creams, eyelash mascara removers, lipstick, nail cuticle products, blushes, soaps, chocolate, creams, ointments, and medicinal suppositories.

Use: As skin softener and lubricant.

Health Effects: A possible allergen.

Synonyms: ✲ THEOBROMA OIL

COCONUT OIL

Products: In margarine, shortenings, deep-frying oil, synthetic cocoa butter, soaps, cosmetics, synthetic detergents, baby soaps, hair products, and cream bases.

Use: Oil for food fats and greases, cleansing products, coating agents, and texturizers.

Health Effects: High fat content. Possible allergen. Harmless when used for intended purpose.

Synonyms: CAS: 8001-31-8 ❊ COPRA OIL ❊ COCONUT PALM OIL ❊ COCONUT BUTTER

COGNAC OIL

Products: In liquors, beverages, ice desserts, candies, bakery goods, and condiments.

Use: As a flavoring agent.

Health Effects: Harmless when used for intended purpose.

Synonyms: ❊ COGNAC OIL, WHITE ❊ COGNAC OIL, GREEN ❊ ETHYL OENANTHATE ❊ WINE YEAST OIL

COPPER ACETORARSENITE

Products: In wood preservative, larvicide, and paints.

Use: For insecticide, wood antifouling paint, and preservative.

Health Effects: Toxic by swallowing.

Synonyms: ❊ CUPRIC ACETOARSENITE ❊ PARIS GREEN ❊ KING'S GREEN ❊ SCHWEINFURT GREEN ❊ IMPERIAL GREEN

CORIANDER OIL

Products: In gin, curry powder, meat, sausage, tooth pastes, raspberry, spice, ginger ale, candies, root beer, and condiments.

Use: As a flavoring agent.

Health Effects: May be irritating to the mouth when eaten. A skin irritant. A possible allergen. FDA states GRAS (generally regarded as safe) when used at moderate levels.

Synonyms: CAS: 8008-52-4 ❊ OIL of CORIANDER

CORN OIL

Products: In bakery products, margarine, mayonnaise, salad oil, leather polish,

Use: As a coating agent, emulsifying agent, texturizing agent, and in

hair dressing, and solvent.

lubricants.

Health Effects: In concentrated amounts it has been found to be a human skin irritant. May be an allergen.

Synonyms: CAS: 8001-30-7 ✲ MAIZE OIL ✲ INDIAN CORN ✲ ZEA MAYS

CORN SYRUP

Products: In candies, artificial fruit toppings, sweet syrups, snack foods, imitation dairy products, and coffee whiteners.

Use: As a sweetener and thickener.

Health Effects: Has no nutritional value and sticks to teeth causing tooth decay. It is usually in low-nutrition, high-calorie foods.

Synonyms: ✲ CORN SYRUP SOLIDS

COTTONSEED OIL

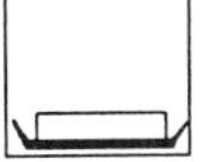

Products: In leather dressing, soap stock, base for cosmetics, nail polish remover, waterproofing products, dietary supplement, candy (hard, and soft chocolate), salad dressings, margarine, mayonnaise, deep-frying oil, lards, and cooking oils.

Use: As a lubricant; various other uses.

Health Effects: Use at a moderate level not in excess of the amount required. Caused tumors and abnormal fetus development in animals. Frequent cause of allergic reactions.

Synonyms: CAS: 8001-29-4 ✲ DEODORIZED WINTERIZED COTTONSEED OIL

COUMARONE-INDENE RESIN

Products: For grapefruit, lemons, limes, oranges, tangelos, tangerines, adhesives, and printing inks.

Use: As a coating (protective).

Health Effects: FDA limits use on weight basis of fruit on which it is used. Note: All fruits should be washed well before using.

Synonyms: None found.

CREOSOTE, COAL-TAR

Products: For railroad ties, wood treatment, foundation coatings,

Use: As wood preservative, biocide, skin medication, and veterinary skin

waterproofing, telephone poles, waterside dock pilings, and fungicide. treatments.

Health Effects: A confirmed carcinogen (causes cancer). Toxic by breathing of fumes. Moderately toxic by ingestion. A skin and eye irritant.

Synonyms: CAS: 8001-58-9 ✲ CREOSOTE OIL ✲ LIQUID PITCH OIL ✲ TAR OIL ✲ BRICK OIL ✲ CRESYLIC CREOSOTE

CROTONALDEHYDE

Products: In insecticide, tear gas, fuel-gas component, and leather tanning.

Use: For pesticides and personal protection, among others.

Health Effects: Irritating to eyes and skin. A lachrymator (causes eyes to water). Flammable. Fire risk.

Synonyms: CAS: 123-73-9 ✲ 2-BUTENAL ✲ CROTONIC ✲ ALDEHYDE

CUMIN OIL

Products: In cheese, meat, relishes, and soups.

Use: For flavoring agent and perfumery.

Health Effects: Moderately toxic by swallowing gross amounts of pure oil. A skin irritant. GRAS (generally regarded as safe) when used at a moderate level.

Synonyms: CAS: 8014-13-9 ✲ CUMMIN ✲ OILS, CUMIN

CUPROUS IODIDE

Products: Dietary supplement added to table salt to prevent goiters.

Use: Source of iodine in table salt.

Health Effects: GRAS (generally regarded as safe) with a limitation of 0.01 percent in table salt.

Synonyms: CAS: 7681-65-4 ✲ COPPER IODIDE ✲ COPPER(I) IODIDE

CYCLAMATE

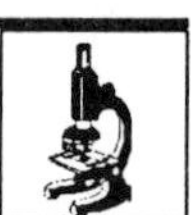

Products: Sweetener (nonnutritive). Approved for use in Europe and in 40 countries around the world.

Use: Artificial sweetener. Currently under consideration for approval in the U. S.

Health Effects: This was a popular sweetener in the 1960's until it was removed from the market in 1969. Suspected human carcinogen (may cause cancer) produced bladder tumors. Mildly toxic by swallowing excessive amounts.

Synonyms: CAS: 100-88-9 ✲ CYCLAMIC ACID ✲ CYCLOHEXANESULPHAMIC ACID ✲ CYCLOHEXYLAMIDOSULPHURIC ACID ✲ CYCLOHEXYLAMINESULPHONIC ACID ✲

CYCLOHEXYLSULFAMIC ACID ✲ CYCLOHEXYLSULPHAMIC ACID ✲ HEXAMIC ACID ✲ SUCARYL ✲ SUCARYL ACID

CYCLOHEXANE

Products: In paint and varnish remover, solid fuels, and fungicides.

Use: For solvents, insecticides, and fungus killers.

Health Effects: Moderately toxic by breathing and skin contact.

Synonyms: CAS: 110-87-7 ✲ HEXAMETHYLENE ✲ HEXANAPHTHENE ✲ HEXALHYDROBENZENE

CYCLOHEXANONE

Products: For wood stains, paint and varnish removers, spot removers, polishes, and lubrication oils.

Use: In solvents, stains, and coatings.

Health Effects: Moderately toxic by swallowing. A skin and severe eye irritant. Toxic effects on the human body by breathing. Can cause changes in the sense of smell, and unspecified respiratory changes. Mild narcotic properties have been ascribed to it. A possible mutagen (may change inherited characteristics). Caused reproductive effects (infertility, or sterility, or birth defects) in animals. Moderate fire risk.

Synonyms: CAS: 108-94-1 ✲ PIMELIC KETONE ✲ KETOHEXAMETHYLENE

p-CYMENE

Products: It is found in nearly 100 volatile oils including lemongrass, sage, thyme, coriander, star anise and cinnamon. In beverages, cream desserts, candy, and bakery goods.

Use: For fragrances, spices, and citrus flavorings for foods.

Health Effects: Pure substance is mildly toxic by swallowing large amounts. Humans sustain central nervous system effects at low doses. A skin irritant. FDA approves use at modest levels to accomplish the intended effect.

Synonyms: CAS: 99-87-6 ✲ CAMPHOGEN ✲ CYMENE ✲ CYMOL ✲ DOLCYMENE ✲ PARACYMENE ✲ PARACYMOL

l-CYSTINE

Products: In baked goods (yeast leavened) and baking mixes.

Use: As a dietary supplement, dough strengthener, and nutrient.

Health Effects: GRAS (generally regarded as safe) when used within FDA limitations.

Synonyms: CAS: 56-89-3 ✲ CYSTEINE DISULFIDE ✲ CYSTIN ✲ (-)-CYSTINE ✲ CYSTINE ACID ✲ DICYSTEINE ✲ β,β'-DITHIODIALANINE ✲ GELUCYSTINE ✲ OXIDIZED l-CYSTEINE

D

"7000 inspectors overlook 100 million animals a year trying to keep diseased meat from the dinner table."
United States Department of Agriculture USDA

2,4-D

Products: On barley (milled fractions, except flour), oats (milled fractions, except flour), potable water, rye (milled fractions), sugarcane bagasse, sugarcane molasses, and wheat (milled fractions, except flour).

Use: As a herbicide, weed killer, and defoliant.

Health Effects: A suspected human carcinogen (may cause cancer). Poison by swallowing. Moderately toxic by skin contact. Effects on the human body by swallowing are: somnolence (sleepiness), convulsions, coma, and nausea or vomiting. Can cause liver and kidney injury. A skin and severe eye irritant. A human mutagen (changes inherited characteristics).

Synonyms: CAS: 94-75-7 ❊ AGROTECT ❊ AQUA-KLEEN ❊ CHLOROXONE ❊ CROP RIDER ❊ DED-WEED ❊ ESTERON BRUSH KILLER ❊ FARMCO ❊ HERBIDAL ❊ LAWN-KEEP ❊ MIRACLE ❊ PLANOTOX ❊ PLANTGARD ❊ SALVO ❊ SUPER D WEEDONE ❊ WEED-B-GON ❊ WEEDEZ WONDER BAR ❊ WEEDONE LV4 ❊ WEED TOX ❊ WEEDTROL

DALAPON

Products: A herbicide.

Use: Kills plants and foliage.

Health Effects: Strong irritant to eyes and skin.

Synonyms: CAS: 75-00-0 ❊ 2,2-DICHLOROPROPIONIC ACID

DDH

Products: In household laundry bleach, water treatments, swimming pool treatments, and mild chlorinating agents.

Use: As a disinfectant, antiseptic, whitening agent, and purifying agent.

Health Effects: Mildly toxic by swallowing and breathing. A severe skin irritant. A mutagen (changes inherited characteristics). Avoid contact because of effects of active chlorine on skin. Sometimes these chemicals are central nervous system depressants (slows heart rate and breathing rate).

Synonyms: CAS: 118-52-5 ✲ DICHLORANTIN ✲ DANTOIN ✲ DACTIN ✲ HYDAN ✲ HALANE ✲ DCA

1-DECANAL

Products: Found in over 50 sources including citrus oils, citronella, and lemongrass. Used as fruit and berry flavoring in beverages, ice desserts, candy, bakery products, gum, and gelatins.

Use: As a flavoring agent.

Health Effects: Moderately toxic by swallowing large amounts of the pure chemical. Mildly toxic by skin contact. A severe skin irritant. Use at a level not in excess of the amount required to accomplish the desired results.

Synonyms: CAS: 112-31-2 ✲ ALDEHYDE C10 ✲ C-10 ALDEHYDE ✲ CAPRALDEHYDE ✲ 1-DECYL ALDEHYDE

DECYL ALCOHOL

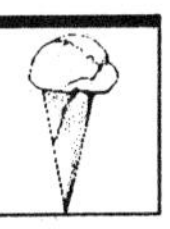

Products: In baked goods, beverages, candy, ice cream, butter, coconut, and fruit. In detergents, lubricants, solvents, and perfumes.

Use: As a flavoring agent.

Health Effects: Moderately toxic by skin contact. The concentrated substance in gross amounts is mildly toxic by swallowing and breathing. A severe skin and eye irritant. Possible carcinogen (may cause cancer) that produced tumors in animals. Use at a level not in excess of the amount required to accomplish the desired results.

Synonyms: CAS: 112-30-1 ✲ ALCOHOL C-10 ✲ ANTAK ✲ CAPRIC ALCOHOL ✲ DECANAL DIMETHYL ACETAL ✲ DECANOL

DECAHYDRONAPHTHALENE

Products: In substitutes for turpentine, stain remover, shoe polish, floor wax, and lubricants.

Use: As a solvent for oils, fats, and resins.

Health Effects: Irritant to eyes and skin.

Synonyms: CAS: 91-17-8 ✲ DECALIN ✲ NAPTHANE ✲ NAPTHALANE

DEET

Products: Insect repellents, solvent, and film former.

Use: Various; primarily as a pesticide.

Health Effects: Toxic by swallowing. Irritant to eyes, nose and throat. Usually 30% concentration or less is effective and not toxic to humans. Avoid using products that contain DEET on cats. Cats are more vulnerable to toxic substances than humans or dogs because they lack the enzymes that break down toxic substances in the liver.

Synonyms: CAS: 134-62-3 ❊ N,N-DIETHYL-m-TOLUAMIDE ❊ DELPHENE ❊ FLYPEL ❊ OFF ❊ REPEL

DEHYDROACETIC ACID

Products: In fungicides, bactericides, medicated toothpaste, toothpowder, makeup, and shampoos.

Use: Destroys fungi and bacteria in dentifrices, cosmetics, hair grooming products, and preservatives.

Health Effects: Toxic by swallowing.

Synonyms: CAS: 520-45-6 ❊ DHA ❊ METHYLACETOPYRANONE ❊ DEHYDRACETIC ACID

DE-ICER CHEMICALS

Products: As flakes, granules or pellets.

Use: To melt ice and snow from walkways or streets. They require moisture to dissolve and form brine solutions that lower the freezing point of water. Their brine solutions melt ice and snow and penetrate down to pavement undercutting the ice from the pavement. Calcium chloride is the only de-icer that gives off heat as it dissolves. This works faster than other types.

Health Effects: Harmless when used according to package directions using appropriate caution.

Synonyms: ❊ CALCIUM CHLORIDE ❊ SODIUM CHLORIDE ❊ ROCK SALT ❊ POTASSIUM CHLORIDE ❊ UREA

DEMETON-S

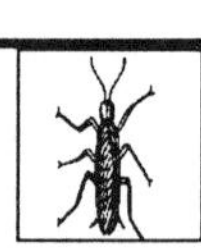

Products: As a pesticide.

Use: A systemic insecticide which is absorbed by the plant which then becomes toxic to sucking and chewing insects.

Health Effects: Poison by swallowing.

Synonyms: CAS: 126-75-0 ❊ DIETHYL-S-(2-ETHIOETHYL)THIOPHOSPHATE ❊ O,O-DIETHYL-S-2-(ETHYLTHIO)ETHYL PHOSPHOROTHIOATE ❊ O,O-DIETHYL-S-(2-(ETHYLTHIO)ETHYL) PHOSPHOROTHIOLATE (USDA) ❊ ISODEMETON ❊ PO-SYSTOX ❊ THIOLDEMETON ❊ SYSTOX

DENATURED ALCOHOL

Products: For solvents, antifreeze, and brake fluids.

Use: Various

Health Effects: Flammable. A dangerous fire risk.

Synonyms: ✲ DENATURED SPIRITS

DEXTRINS

Products: In baked goods, beverages (dry mix), confectionery products, egg roll, food-contact surfaces, gravies, pie fillings, poultry, puddings, and soups.

Use: As a binder, stabilizer, extender, adhesives, printing inks, surface-finishing agent for paper, textiles, and a thickening agent.

Health Effects: GRAS (generally regarded as safe) when used at moderate levels for intended purposes. Possible allergen.

Synonyms: CAS: 9004-53-9 ✲ ARTIFICIAL GUM ✲ DEXTRANS ✲ STARCH GUM ✲ TAPIOCA ✲ VEGETABLE GUM

DIALIFOR

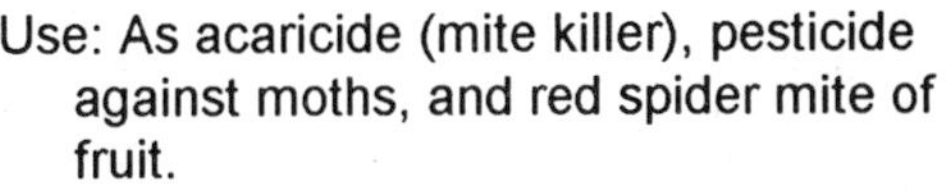

Products: Insecticide for apple pomace (dried), citrus pulp (dried), grape pomace (dried), raisin waste, and raisins.

Use: As acaricide (mite killer), pesticide against moths, and red spider mite of fruit.

Health Effects: Poison by swallowing and skin contact. An animal teratogen (abnormal fetus development). Other animal reproductive effects (infertility, or sterility, or birth defects). FDA states specific limitations for use on food products.

Synonyms: CAS: 10311-84-9 ✲ S-(2-CHLORO-1-PHTHALIMIDOETHYL)-O,O-DIETHYL PHOSPHORODITHIOATE ✲ O,O-DIETHYL-S-(2-CHLORO-1-PHTHALIMIDOETHYL)PHOSPHORODITHIOATE ✲ TORAK

DIAMYL PHENOL

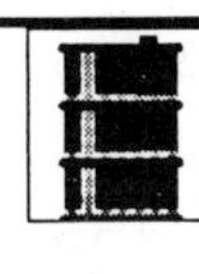

Products: In oils, rust preventatives, and detergents.

Use: For lubrication and cleaning.

Health Effects: Irritant to skin. Could cause allergic reaction.

Synonyms: ✲ 1-HYDROXY-2,4-DIAMYLBENZENE

DIATOMACEOUS EARTH

Products: In paints, tooth polish, cosmetics, and facial powder.

Use: As abrasive, absorbent, filter aid, and anticaking agents.

Health Effects: A nuisance dust which may cause fibrosis of the lungs. A possible carcinogen (may cause cancer). Composed of the skeletons of small aquatic plants.

Synonyms: CAS: 61790-53-2 ✲ D.E. ✲ DIATOMACEOUS SILICA ✲ DIATOMITE ✲ INFUSORIAL EARTH ✲ KIESELGUHR

DIAZINON

Products: Insecticide.

Use: As a pesticide that is particularly effective against fire ants. This use permitted by EPA.

Health Effects: Poison by swallowing and skin contact. Mildly toxic by breathing. Effects on the body by swallowing are: changes in motor activity, muscle weakness, and sweating. Animal teratogenic (abnormal fetus development) and reproductive effects (infertility, or sterility, or birth defects). A skin and severe eye irritant. A human mutagen (changes inherited characteristics).

Synonyms: CAS: 333-41-5 ✲ ALFA-TOX ✲ DIANON ✲ DIAZIDE ✲ DIAZINONE ✲ DIAZITOL ✲ DIAZOL ✲ DIZINON ✲ GARDENTOX ✲ SPECTRACIDE

DIBENZYL DISULFIDE

Products: In petroleum oils and greases.

Use: Lubricant.

Health Effects: Moderately toxic by swallowing. A skin and eye irritant.

Synonyms: CAS: 150-60-7 ✲ BENZYL DISULFIDE

DIBENZYL ETHER

Products: In beverages, ice desserts, candy, bakery goods, and gum.

Use: As a synthetic spice and fruit flavoring.

Health Effects: Pure chemical is moderately toxic by swallowing gross amounts. Vapors are probably narcotic in high concentration. A skin and eye irritant.

Synonyms: CAS: 103-50-4 ✲ BENZYL ETHER

DIBROMODIFLUOROMETHANE

Products: Fire extinguishing agent and direct-contact freezing agent.

Use: To extinguish flames and to freeze food products.

Health Effects: Mildly toxic by breathing.

Synonyms: CAS: 75-61-6 ✲ DIFLUORODIBROMOMETHANE ✲ FREON 12-B2 ✲ HALON 1202

DIBROMOFLUORESCEIN

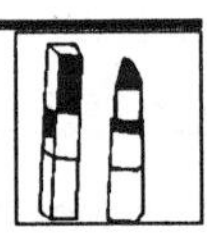

Products: In lipstick and blushes.

Use: As a red-orange coloring or dye.

Health Effects: A possible allergen and stomach irritant upon swallowing. Possible skin irritant.

Synonyms: None found.

DIBROMOPROPANOL

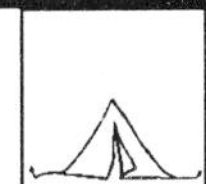

Products: In flame retardant products.

Use: As a coating.

Health Effects: A carcinogen and possible mutagen (changes inherited characteristics).

Synonyms: CAS: 96-13-9 ✲ 2,3-DIBROMO-1-PROPANOL

DIBUTYL BUTYL PHOSPHONATE

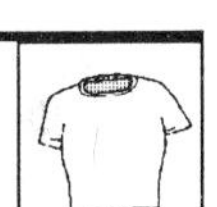

Products: Fabric softener.

Use: As a textile conditioner and anti-static agent.

Health Effects: A possible allergen.

Synonyms: None found.

DIBUTYL PHTHALATE

Products: In cosmetics, inks, perfumes, and glues.

Use: As a plasticizer in nail polish, perfume ingredient (fixative), printing inks, paper coatings, adhesives, and insect repellent for textiles.

Health Effects: Mildly toxic by swallowing. Effects on the human body by swallowing are; hallucinations, distorted perceptions, nausea or vomiting, kidney, ureter, or bladder changes. A possible mutagen (changes inherited characteristics).

Synonyms: CAS: 84-74-2 ✲ DBP ✲ CELLUFLEX ✲ PALATINOL ✲ 1,2-BENZENE-DICARBOXYLATE

DIBUTYL SEBACATE

Products: Plasticizer and fruit flavored cosmetics and perfumes.

Use: In packaging materials, odorants, and flavorings.

Health Effects: Mildly toxic by swallowing. Animal reproductive effects (infertility, or sterility, or birth defects).

Synonyms: CAS: 109-43-3 ✲ DECANEDIOIC ACID, DIBUTYL ESTER ✲ DI-n-BUTYL SEBACATE ✲ KODAFLEX DBS ✲ MONOPLEX DBS ✲ POLYCIZER DBS ✲ SEBACIC ACID, DIBUTYL ESTER ✲ STAFLEX DBS

DICHLOROBENZALKONIUM CHLORIDE

Products: In agricultural products.

Use: As algicide (kills algae), antiseptic, and sterilizing agent.

Health Effects: A deadly poison by swallowing. A severe eye irritant. Can cause liver and kidney damage. Could cause allergic reaction.

Synonyms: CAS: 8023-53-8 ✲ TETROSAN

o-DICHLOROBENZENE

Products: In fumigants, insecticides, vehicle polishes, air deodorants (fresheners), shoe dyes, shoe polishes, tar removers, and grease removers.

Use: As a pesticide, polish, odorant, and solvent.

Health Effects: Poison by swallowing. Moderately toxic by breathing. An animal teratogen (abnormal fetus development), which also caused reproductive effects (infertility, or sterility, or birth defects). An eye, skin, nose, and throat irritant. A possible carcinogen (may cause cancer). A mutagen (changes inherited characteristics).

Synonyms: CAS: 95-50-1 ✲ 1,2-DICHLOROBENZENE ✲ CHLOROBEN ✲ CHLORODEN ✲ DCB ✲ TERMITKIL

p-DICHLOROBENZENE

Products: As a fumigant, moth repellent, germicide, space odorant, and soil fumigant.

Use: As a pesticide, bactericide, and air deodorants.

Health Effects: A definite carcinogen (causes cancer). An animal teratogen (abnormal fetus development). Moderately toxic to humans by swallowing. Various effects on the body by swallowing include liver changes, respiratory effects, and constipation. An eye irritant.

Synonyms: CAS: 106-46-7 ✲ 1,4-DICHLOROBENZENE ✲ PARACIDE ✲ PARADICHLORBENZENE ✲ PARAMOTH ✲ PARAZENE ✲ PDB ✲ PDBC

DICHLORAMINE-T

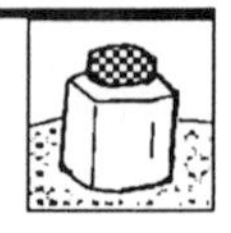

Products: In antiseptic medication.

Use: A germicide and antibacterial.

Health Effects: A possible allergen.

Synonyms: ✲ CHLORAMINE-T ✲ SODIUM p-TOLUENESULFOCHLOR ✲ p-TOLUENESULFONDICHLORAMIDE

DICHLORODIFLUOROMETHANE

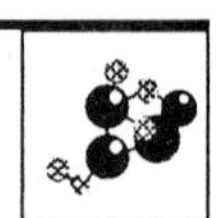

Products: A refrigerant in air conditioners for freezing of foods by direct contact, and the chilling of cocktail glasses.

Use: For direct contact freezing agent.

Health Effects: Effects on the body by breathing include eye irritation, lung irritation, and liver changes. A narcotic in high concentrations.

Synonyms: CAS: 75-71-8 ❖ FLUOROCARBON 12 ❖ ARCTON ❖ FREON F 12 ❖ PROPELLANT 12 ❖ HALON

DICHLOROETHER

Products: In paints, varnishes, lacquers, finish removers, spot remover, and drycleaning fluid.

Use: As a solvent and in penetrating compounds.

Health Effects: A poison by swallowing, skin contact, and breathing. A skin, eye, nose, and throat irritant. Questionable carcinogen (may cause cancer).

Synonyms: CAS: 111-44-4 ❖ DICHLOROETHYL ETHER ❖ CHLOREX ❖ DCEE

α-DICHLOROHYDRIN

Products: General solvent for paints, varnishes, and lacquers.

Use: As a dissolver and binder in paint products.

Health Effects: Toxic by breathing and swallowing.

Synonyms: ❖ DICHLOROISPROPYL ALCOHOL ❖ 1,3-DICHLORO-2-PROPANOL

DICHLOROISOCYANURIC ACID

Products: In household dry bleaches, dishwashing compounds, scouring powders, and detergent sanitizers.

Use: As a replacement for calcium hypochlorite.

Health Effects: May ignite organic materials on contact. Irritant to eyes.

Synonyms: ❖ DICHLORO-S-TRIAZINE-2,4,6-TRIONE

DICHLOROISOPROPYL ETHER

Products: In spot remover, dry cleaning solutions, paint, and varnish remover.

Use: As a solvent for oils, greases, fats, and waxes.

Health Effects: Moderately toxic by swallowing. Moderately toxic by skin contact and breathing. An eye irritant. Questionable carcinogen (may cause cancer). A corrosive material.

Synonyms: CAS: 108-60-1 ❖ BIS(2-CHLORO-1-METHLETHYL)ETHER ❖ DICHLORODIISOPROPYL ETHER

DICHLOROPENTANE

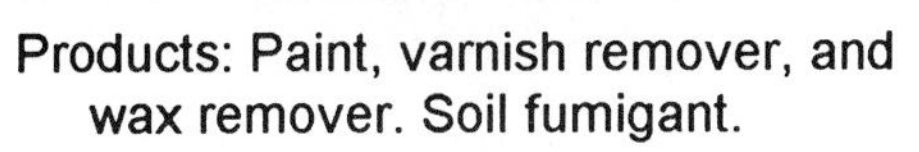

Products: Paint, varnish remover, and wax remover. Soil fumigant.

Use: As a solvent for oil, grease, and tar.

Health Effects: Flammable, dangerous fire risk.

Synonyms: CAS: 30586-10-8 ✲ CHLORINATED HYDROCARBONS, ALIPHATIC

DICHLOROPHENE

Products: In cosmetics, deodorants, shampoos, and toothpaste.

Use: As a fungicide and germ killer.

Health Effects: Moderately toxic by swallowing. A skin and severe eye irritant. Frequently causes allergic reaction. Can cause cramps and diarrhea.

Synonyms: CAS: 97-23-4 ✲ ANTIPHEN ✲ PANACIDE ✲ TENIATHANE ✲ WESPURIAL

DICHLORVOS

Products: In flea collars and sprays. In roach and ant killers. For animal feed, cereals, cookies (packaged), crackers (packaged), figs (dried), flour, pork, and sugar.

Use: Insecticide.

Health Effects: Poison by swallowing gross amounts of pure chemical. Also poison by breathing and skin contact. In animals it caused teratogenic (abnormal fetus development) and reproductive effects (infertility, or sterility, or birth defects). A human mutation (changes inherited characteristics) data reported. It is used in flea and tick collars for dogs. FDA states limits on amounts used on human foods.

Synonyms: CAS: 62-73-7 ✲ CANOGARD ✲ CHLORVINPHOS ✲ CYANOPHOS ✲ DDVF ✲ DDVP ✲ DERIBAN ✲ DICHLOROPHOS ✲ FLY-DIE ✲ NO-PEST ✲ NO-PEST STRIP ✲ TASK TABS

DICYCLOHEXYLAMINE

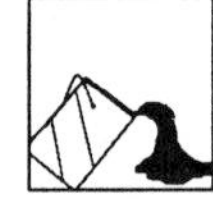

Products: In paints, varnishes, inks, and detergents.

Use: Various.

Health Effects: Toxic by swallowing, strong irritant to skin, nose, and throats.

Synonyms: CAS: 101-83-7 ✲ CDHA ✲ DODECAHYDRODOPHENYLAMINE

DIETHANOLAMINE

Products: In liquid detergents, shampoos, cleaners, and polishes.

Use: As a dispersing agent, emollient (softener), humectant (collects moisture).

Health Effects: A poison by swallowing, skin contact, and breathing. A skin, eye, nose, and throat irritant. A possible carcinogen (may cause cancer).

Synonyms: CAS: 111-42-2 ✲ DEA ✲ DIOLAMINE ✲ DIETHYLOLAMINE

DIETHYLAMINOETHANOL

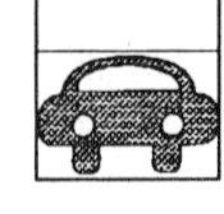

Products: In fabric softeners and anti-rust compositions.

Use: Various.

Health Effects: Toxic by swallowing and skin absorption.

Synonyms: CAS: 100-37-8 ✲ DIETHYLETHANOLAMINE ✲ 2-HYDROXYTRIETHYLAMINE

DIETHYL DICARBONATE

Products: Inhibits fermentation.

Use: As a fungicide.

Health Effects: Poison by swallowing. Concentrated DEPC is irritating to eyes, nose, throats, and skin. Prohibited from direct addition or use in human food. Legal for use in wine in other countries.

Synonyms: CAS: 1609-47-8 ✲ BAYCOVIN ✲ DEPC ✲ DICARBONIC ACID DIETHYL ESTER ✲ DIETHYL ESTER of PYROCARBONIC ACID ✲ DIETHYL OXYDIFORMATE ✲ DIETHYL PYROCARBONIC ACID

DIETHYLENE ETHER

Products: In lacquer, paint, varnish, solvents and removers. Cleaning preparations, detergent preparations, cements, cosmetics, and deodorants.

Use: Various.

Health Effects: Toxic by breathing. Absorbed by skin. A carcinogen (causes cancer).

Synonyms: CAS: 123-91-1 ✲ 1,4-DIOXANE ✲ DIETHYLENE OXIDE ✲ DIOXYETHYLENE ETHER ✲ 1,4-DIETHYLENE DIOXIDE

DIETHYLENE GLYCOL

Products: In fabric softeners, tobacco moisturizer, synthetic sponges, paper products, corks, book-binding adhesives, cosmetics, rug cleaner, upholstery cleaners, floor polish, furniture polish, disinfectants, and antifreeze.

Use: As a conditioner, humectant (moisturizer), and cleaner.

Health Effects: Moderately toxic to humans by swallowing. A skin and eye irritant.

Synonyms: CAS: 111-46-6 ✲ DIHYDROXYDIETHYL ETHER ✲ DIGLYCOL ✲ DEG ✲ GLYCOL ETHYL ETHER

DIETHYL PHTHALATE

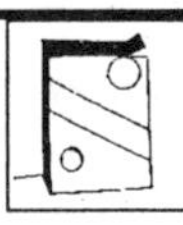

Products: A plasticizer (for flexibility) in wrapping materials. Substance is also found in mosquito repellent.

Use: In packaging materials and insecticidal sprays.

Health Effects: Moderately toxic by swallowing. Effects by breathing are: lachrymation, (watering eyes), respiratory obstruction, and other unspecified respiratory system effects. An eye irritant. Narcotic in high concentrations.

Synonyms: CAS: 84-66-2 ✲ ANOZOL ✲ DIETHYL-o-PHTHALATE ✲ ETHYL PHTHALATE ✲ NEANTINE ✲ PHTHALOL ✲ SOLVANOL

DIFLUORPHOSPHORIC ACID

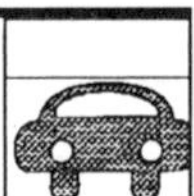

Products: In polish for vehicle surfaces and metal protectors. .

Use: As protective coating for metal.

Health Effects: Could be corrosive to skin or eyes. An irritant.

Synonyms: ✲ PHOSPHORODIFLUORIDIC ACID

DIGITOXIN

Products: As prescription medication.

Use: For cardiac treatment.

Health Effects: Derived from the *Foxglove* plant. Can be toxic by swallowing, overdose can be fatal.

Synonyms: CAS: 71-63-6

DIGLYCOL LAURATE

Products: In hand lotions, hair dressings, and dry cleaning soap.

Use: As an emulsifier (to aid in suspension of oily liquids). Prevents separation.

Health Effects: Can cause allergic reactions.

Synonyms: ✲ DIETHYLENE GLYCOL MONOLAURATE

DIGLYCOL MONOSTEARATE

Products: In skin moisturizers, lotions, creams, cosmetics, and makeup.

Use: As an emulsifier (to aid in suspension of oily liquids) and as a thickener.

Health Effects: Can cause allergic reactions.

Synonyms: CAS: 106-11-6 ✲ DIETHYLENE GLYCOL MONOSTEARATE

DIGLYCOL STEARATE

Products: In powders, polishes, and cleaners.

Use: As an emulsifier (to aid in suspension of oils), as a lubricant, and as a thickener or filler.

Health Effects: Can cause allergic reactions.

Synonyms: ✲ DIETHYLENE GLYCOL DISTEARATE

DIHYDROXYACETONE

Products: In cosmetic products used to produce artificial suntan appearance on skin.

Use: As skin makeup.

Health Effects: Possible skin irritant and possible cause of allergic reactions.

Synonyms: ✲ DHA ✲ DIHYDROXYPROPANONE

2,5-DIHYDROXYBENZOQUINONE

Products: In skin products.

Use: As a tanning agent.

Health Effects: Derived from hydroquinone. An irritant to skin and eyes.

Synonyms: None found.

5,7-DIHYDROXY-4-METHYLCOUMARIN

Products: In suntan oils and lotion products. In wall paints as a whitening agent.

Use: As a sunscreen. As a pigment in paints. An UV (ultraviolet) light absorber.

Health Effects: Possible cause of allergic reactions.

Synonyms: None found.

DIHYDROXYSTEARIC ACID

Products: In cosmetics and lotions.

Use: As a thickener and stabilizer in makeup products.

Health Effects: Can cause allergic reactions.

Synonyms: None found.

DIISOBUTYL KETONE

Products: In lacquers, inks, and stains.

Use: As a solvent in paint-type products.

Health Effects: Moderately toxic by swallowing and breathing. Mildly toxic by skin contact. An eye and skin irritant. Narcotic in high concentration. breathing can cause headache, nausea, and vomiting.

Synonyms: CAS: 108-83-8 ⁂ ISOVALERONE ⁂ VALERONE

DIISOPROPANOLAMINE

Products: In polishes, leather preservatives, leather polishes, oils, and water paints.

Use: An emulsifier (aids in suspension of oils), stabilizer, and maintains mixes of chemicals in products.

Health Effects: Can cause allergic reactions.

Synonyms: ⁂ DIPA

DIISOPROPYL DIXANTHOGEN

Products: In fungicides and weed killers.

Use: An herbicide.

Health Effects: Toxic by swallowing and breathing. A strong irritant.

Synonyms: None found.

DIKETENE

Products: A food preservative.

Use: To maintain food quality.

Health Effects: In large amounts it is moderately toxic by swallowing pure substance. A skin and severe eye irritant.

Synonyms: CAS: 674-82-8 ⁂ ACETYL KETENE ⁂ KETENE DIMER

DILL SEED OIL, INDIAN TYPE

Products: In dips, meats, sauces, and spreads.

Use: As a flavoring agent.

Health Effects: GRAS (generally regarded as safe).

Synonyms: ⁂ DILL OIL, INDIAN TYPE ⁂ DILL SEED OIL, INDIAN

DILL WEED OIL

Products: In bakery goods, sauces, meat sausages, and pickles.

Use: As a seasoning.

Health Effects: In large amounts it could by mildly toxic by swallowing pure substance. A skin irritant.

Synonyms: CAS: 8006-75-5 ✲ DILL FRUIT OIL ✲ DILL HERB OIL ✲ DILL OIL ✲ DILL SEED OIL ✲ DILL WEED OIL

DIMETHOXANE

Products: Preservative for cosmetics and inks. A gasoline additive.

Use: A chemical that stabilizes and preserves products.

Health Effects: Possible carcinogen (may cause cancer). Moderately toxic by swallowing.

Synonyms: CAS: 828-00-2 ✲ ACETOMETHOXANE ✲ DIOXIN BACTERICIDE

N,N-DIMETHYL ACETAMIDE

Products: In paint remover.

Use: A solvent.

Health Effects: Toxic by breathing. It is absorbed by skin. A strong irritant.

,Synonyms: CAS: 127-19-5 ✲ DMAC

DIMETHYLAMINOETHYL METHACRYLATE

Products: A fabric antistatic agent.

Use: As a fabric softener or coating agent for textiles.

Health Effects: An irritant to skin, eyes, nose, and throats. A strong lachrymator (causes eyes to water).

Synonyms: None found.

2,5-DIMETHYLBENZYL CHLORIDE

Products: In dyes, perfumes, and germicides.

Use: Various.

Health Effects: Irritant to eyes, nose, and throats. A lachrymator (causes eyes to water).

Synonyms: ✲ α-CHLORO-p-XYLENE

DIMETHYL CHLOROTHIOPHOSPHATE

Products: In flame retardant coatings, sprays, fungicides, and pesticides.

Use: For camping equipment and tent fabric coatings, insecticides, and additives.

Health Effects: Poison by breathing. Moderately toxic by swallowing and skin contact. Corrosive. Use may be restricted.

Synonyms: CAS: 2524-03-0 ✲ METHYL PCT ✲ DIMETHYL PHOSPHOROCHLORIDOTHIOATE

DIMETHYL DICARBONATE

Products: In wine. | Use: As fungicide and yeast inhibitor.

Health Effects: FDA requires limitation of 200 ppm in wine.

Synonyms: CAS: 4525-33-1

DIMETHYLHYDANTOIN-FORMALDEHYDE POLYMER

Products: In adhesives and aerosol hair sprays. | Use: As a fixative.

Health Effects: Avoid breathing.

Synonyms: None found.

DIMETHYLOCTANOL

Products: In bakery products, beverages (nonalcoholic), chewing gum, confections, ice cream products, and pickles. | Use: As a floral and fruit flavoring agent.

Health Effects: Moderately toxic by skin contact of concentrated chemical. A skin irritant.

Synonyms: CAS: 106-21-8 ✲ DIHYDROCITRONELLOL ✲ GERANIOL TETRAHYDRIDE ✲ PELARGOL ✲ PERHYDROGERANIOL ✲ TETRAHYDROGERANIOL

2,6-DIMETHYLMORPHOLINE

Products: In rubless floor polishes, germicides, and textile finishes. | Use: Various.

Health Effects: Moderately toxic by swallowing and skin contact. A skin irritant.

Synonyms: CAS: 141-91-3

DIMETHYLOLUREA

Products: Resin used in the formation of plywood and pressboard lamination. | Use: To increase the hardness and fire resistance in wood products.

Health Effects: Avoid breathing of vapors as degassing occurs. Ventilate area where the wood and board is installed.

Synonyms: ✲ DMU ✲ 1,3-BISHYDROXYMETHLUREA

DIMETHYL PHTHALATE

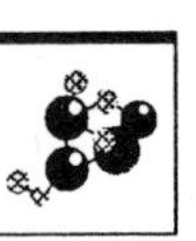

Products: In lacquers, plastics, and rubber | Use: A coating agent to aid flexibility.

products.

Health Effects: Moderately toxic by swallowing. Mildly toxic by breathing. An irritant to eyes, nose, and throats. Not absorbed by the skin.

Synonyms: CAS: 131-11-3 ❖ AVOLIN ❖ DMP ❖ METHYL PHTHALATE ❖ PHTHALIC ACID METHYL ESTER

DIMETHYLPOLYSILOXANE

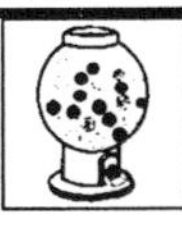

Products: In chewing gum, gelatins, poultry, salt, sugar, wine, transformer liquid, and brake fluids.

Use: As a defoamer and release agent.

Health Effects: FDA approves use within limitations.

Synonyms: ❖ DIMETHYL SILICONE ❖ POLYDIMETHYLSILOXANE

DIMETHYL SULFIDE

Products: Gas odorant.

Use: To give natural gas an identifiable odor.

Health Effects: Flammable, dangerous fire risk, moderate explosion risk.

Synonyms: CAS: 75-18-3 ❖ METHYL SULFIDE

DIMETHYL SULFOXIDE

Products: In industrial cleaners, pesticides, and paint stripping.

Use: Solvent. It is believed by some that it is an effective anti-inflammatory for arthritic conditions.

Health Effects: Poison by swallowing. A skin and eye irritant. A human mutagen (changes inherited characteristics). Readily penetrates skin and other tissues.

Synonyms: CAS: 67-68-5 ❖ DMSO ❖ METHYL SULFOXIDE

DIMYRISTYL ETHER

Products: In water repellent sprays and coatings. As antistatic fabric softeners.

Use: As coating or conditioner.

Health Effects: Could cause allergic reactions on skin contact.

Synonyms: ❖ DITETRADECYL ETHER

2,4-DINITROANISOLE

Products: In moth, furniture, carpet beetle, cockroach, and body lice

Use: As a pesticide.

repellent. In insecticides.

Health Effects: Toxic by swallowing.

Synonyms: ❖ 2,4-DINITRIOPHENYL METHYL

DINITROPHENOL

Products: In dyes, lumber preservative, and photographic developer.

Use: Various.

Health Effects: Absorbed by skin. Dust breathing may be fatal.

Synonyms: CAS: 25550-58-7

DINOSEB

Products: A herbicide for preemergence treatment.

Use: Agricultural. Prevents the sprouting of weeds.

Health Effects: Possible fire risk. Absorbed by skin. A strong irritant.

Synonyms: CAS: 88-85-7 ❖ 2-SEC-BUTYL-4,6-DINITROPHENOL ❖ 2,4-DINITRO-6-SEC-BUTYLPHENOL

DIPENTENE

Products: In paints, enamels, lacquers, varnishes, printing inks, perfumes, flavors, floor waxes, and furniture polishes.

Use: As a solvent or wax.

Health Effects: A skin irritant.

Synonyms: CAS: 138-86-3 ❖ CINENE ❖ LIMONENE, INACTIVE ❖ CAJEPUTENE ❖ DIPANOL ❖ KAUTSCHIN

DIPHENYLAMINE CHLOROARSINE

Products: Poison gas, and wood treatment products.

Use: For self defense, crowd control gas, or a preservative.

Health Effects: Toxic by breathing and swallowing. A strong irritant.

Synonyms: CAS: 578-94-9 ❖ ADAMSITE ❖ PHENARSAZINE CHLORIDE ❖ DM

DIPHENYLMETHANE

Products: Dyes and perfumery.

Use: Various.

Health Effects: Toxic by swallowing.

Synonyms: ❖ BENZYLBENZENE

DIPHENYL OXIDE

Products: For perfumes particularly in soaps.

Use: In fragrances.

Health Effects: Toxic by inhaling concentrated vapor.

Synonyms: CAS: 101-84-8 ❊ PHENYL ETHER ❊ DIPHENYL ETHER

DIPOTASSIUM PERSULFATE

Products: For fresh citrus fruit juice.

Use: As a defoaming agent.

Health Effects: The pure substance is moderately toxic in gross amounts. An irritant and allergen. FDA approves use at limited levels.

Synonyms: CAS: 7727-21-1 ❊ POTASSIUM PEROXYDISULFATE ❊ POTASSIUM PERSULFATE

DIPROPYLENE GLYCOL METHYL ETHER

Products: In vehicle solvents and brake fluids.

Use: Various.

Health Effects: Mildly toxic by swallowing and skin contact. A mild allergen. A skin and eye irritant.

Synonyms: CAS: 34590-94-8 ❊ DIPROPYLENE GLYCOL MONOMETHYL ETHER

DIPROPYLENE GLYCOL MONOSALICYLATE

Products: In sunscreening products and protective coatings.

Use: As a UV (ultraviolet) light screening agent.

Health Effects: Could cause allergic reaction.

Synonyms: ❊ DIPROPYLENE GLYCOL MONOESTER ❊ SALICYLIC ACID DIPROPYLENE GLYCOL MONOESTER

DIQUAT

Products: Herbicide and defoliant.

Use: A plant growth regulator and suppressant.

Health Effects: Poison by swallowing. A skin and eye irritant. A human mutagen (changes inherited characteristics). FDA limits use.

Synonyms: CAS: 85-00-7 ❊ AQUACIDE ❊ DEIQUAT ❊ DEXTRONE ❊ DIQUAT DIBROMIDE ❊ FEGLOX ❊ PREEGLONE ❊ REGLON ❊ WEEDTRINE-D

DISODIUM EDTA

Products: In salad dressings, margarines, mayonnaise, spreads, processed vegetables, fruits, soft drinks, and canned shellfish.

Use: As a preservative in foods. In medicine it is used as a chelating agent for treatment of lead poisoning.

Health Effects: Moderately toxic by swallowing large amounts of pure substance. Caused reproductive (infertility, or sterility, or birth defects) teratogenic (abnormal fetal development) and mutagenic (changes inherited characteristics) in animal studies.

Synonyms: CAS: 139-33-3 ✲ ETHYLENEDIAMINE TETRAACETIC ACID ✲ DISODIUM SALT ✲ DISODIUM TETRACEMATE ✲ DISODIUM VERSENATE ✲ DISODIUM EDATHAMIL

DISODIUM INOSINATE

Products: For ham, meat (cured), poultry, and sausage.

Use: As a flavor enhancer.

Health Effects: FDA approves use at levels to accomplish the intended effect. Concentrated chemical is moderately toxic in excessive amounts.

Synonyms: CAS: 4691-65-0 ✲ DISODIUM IMP ✲ DISODIUM-5′-INOSINATE ✲ DISODIUM INOSINE-5′-MONOPHOSPHATE ✲ DISODIUM INOSINE-5′-PHOSPHATE

DISODIUM METHYLARSONATE

Products: As a defoliant (crabgrass killer).

Use: A herbicide.

Health Effects: Toxic by swallowing and breathing.

Synonyms: CAS: 144-21-8 ✲ DMA ✲ DISODIUM METHANEARSONATE ✲ METHANEARSONIC ACID ✲ DISODIUM SALT

DISODIUM PYROPHOSPHATE

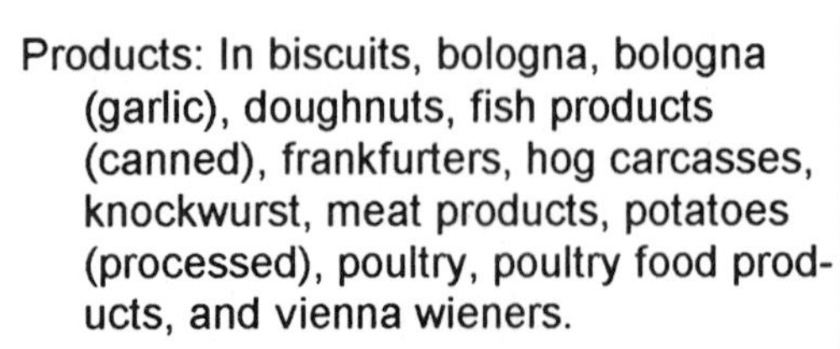

Products: In biscuits, bologna, bologna (garlic), doughnuts, fish products (canned), frankfurters, hog carcasses, knockwurst, meat products, potatoes (processed), poultry, poultry food products, and vienna wieners.

Use: As a sequestrant (binds constituents that affect the final products appearance, flavor or texture), an emulsifier (stabilizes and maintains mixes), and as a texturizer.

Health Effects: GRAS (generally regarded as safe) by FDA when used within limitations.

Synonyms: CAS: 7758-16-9 ✲ DIPHOSPHORIC ACID, DISODIUM SALT ✲ DISODIUM DIHYDROGEN PYROPHOSPHATE ✲ DISODIUM DIPHOSPHATE ✲ SODIUM ACID PYROPHOSPHATE ✲ SODIUM PYROPHOSPHATE

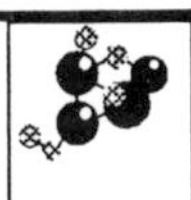

DODECARBONIUM CHLORIDE

Products: Various assorted items.

Use: As a biocide and a disinfectant.

Health Effects: Toxic by swallowing.

Synonyms: None found.

DODECENE

Products: In flavors, perfumes, medicine, and oils.

Use: Various assorted.

Health Effects: Irritant and narcotic in high concentrations.

Synonyms: CAS: 6842-15-5 ❊ 1-DODECENE ❊ DODECYLENE ❊ PROPENE TETRAMER ❊ TETRAPROPYLENE

DODECYL ALCOHOL

Products: In detergents and perfumes.

Use: As a scenting agent. An ingredient that affects the smell of product.

Health Effects: Mildly toxic by swallowing. A severe human skin irritant. Possible human carcinogen (may cause cancer).

Synonyms: CAS: 112-53-8 ❊ ALCOHOL C-12 ❊ ALFOL 12 ❊ CACHALOT ❊ n-DODECANOL ❊ 1-DODECANOL ❊ n-DODECYL ALCOHOL ❊ DUODECYL ALCOHOL ❊ LAURIC ALCOHOL ❊ LAURYL ALCOHOL

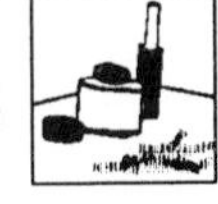

DODECYL BENZENE SODIUM SULFONATE

Products: In cosmetic ingredients and detergents.

Use: Various.

Health Effects: Moderately toxic by swallowing. A skin and severe eye irritant.

Synonyms: CAS: 25155-30-0 ❊ DETERGENT ALKYLATE ❊ BIO-SOFT ❊ DDBSA ❊ DODECYLBENZENESULFONIC ACID SODIUM SALT ❊ DODECYLBENZENESULPHONATE, SODIUM SALT ❊ SODIUM DODECYLBENZENESULFONATE ❊ SODIUM LAURYLBENZENESULFONATE ❊ SULFRAMIN ❊ ULTRAWET

DODECYLBENZYL MERCAPTAN

Products: Metal compound and an odorant chemical.

Use: For cleaning and polishing metals. An odorant for warning of natural gas leaks.

Health Effects: An irritant.

Synonyms: None found.

DODECYL GALLATE

Products: In cream cheese, fats, margarine, oil, oleomargarine, and potatoes (instant mashed).

Use: An antioxidant (slows down the spoiling of fats due to oxidation).

Health Effects: FDA approves use within limitations.

Synonyms: CAS: 1166-52-5 ✲ LAURYL GALLATE ✲ NIPAGALLIN LA ✲ PROGALLIN LA

DODECYL TRIMETHYLAMMONIUM CHLORIDE

Products: In germicides and fungicides.

Use: As a fabric softener and mildew preventer.

Health Effects: Could cause allergic reactions.

Synonyms: None found.

> "Environmental illness, a disorder caused by low doses of everyday chemicals, is increasingly being applied as a diagnosis."
>
> *Vegetarian Times Magazine*

EOSIN

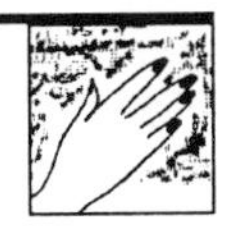

Products: In red ink, cosmetic products, lipstick, nail polish, and motor fuel coloring.

Use: As a dye and coloring agent.

Health Effects: FDA approves use in drugs and cosmetics except for use in eye area for eye makeup. Could cause an allergic reaction.

Synonyms: CAS: 15086-94-9 ✲ BROMEOSIN ✲ TETRABROMOFLUORESCEIN

EPICHLOROHYDRIN

Products: In paints, varnishes, nail polishes, enamels, and lacquers.

Use: As a solvent and for manufacturing process.

Health Effects: Toxic by breathing, swallowing, and skin absorption; strong irritant. A carcinogen (causes cancer).

Synonyms: CAS: 106-89-8 ✲ CHLOROPROPYLENE OXIDE ✲ CHLOROMETHYLOXIRANE

EPOXY RESIN

Products: For floor surfacing, wall panels, cements, mortars, nonskid road surfacing, foams, matrix for stained glass windows, and household appliances.

Use: As a surface-coating for various items.

Health Effects: A strong skin irritant in uncured state.

Synonyms: None found.

ERGOSTEROL

Products: In yeast and mold; widely distributed in nature.

Use: When exposed to ultraviolet light it is converted to vitamin D, an antirachitic (promotes cure, prevents development of Rickets).

Health Effects: Overuse may be harmful due to its ability to catalyze calcium deposits in the bone.

Synonyms: PROVITAMINE D_2

ERGOT

Products: A fungus. Source of many alkaloids and medicines

Use: For pharmaceuticals in appropriate doses. A vasoconstrictor.

Health Effects: A poison by swallowing. Effects on the body by swallowing include nervousness, diarrhea, cyanosis (blue skin from lack of oxygen), excessive thirst, heart rhythm problems, gangrene, and circulatory changes.

Synonyms: 129-51-1 ✲ SECALE CORNUTUM ✲ RYE ERGOT ✲ RYE SCUM ✲ CRUDE ERGOT ✲ CORNOCENTIN ✲ ERGOTRATE

ERYTHORBIC ACID

Products: In bananas (frozen), beef (cured), cured meat food product, pork (cured), pork (fresh), and poultry.

Use: As as antioxidant (slows down spoiling), curing accelerator, and preservative.

Health Effects: FDA states GRAS (generally regarded as safe) when used within limitations.

Synonyms: *d*-ARABOASCORBIC ACID

ERYTHRITOL ANHYDRIDE

Products: As a biocide or bacteriostat.

Use: In germicides and cleansers.

Health Effects: Quite toxic.

Synonyms: CAS: 564-00-1 ✲ BUTADIENE DIOXIDE

ERYTHROSINE

Products: In foods.

Use: For food colors and stains.

Health Effects: In large amounts it is moderately toxic by swallowing concentrated chemical. A possible carcinogen (may cause cancer). A mutagen (changes inherited characteristics). FDA permits use only after certification. Use is then approved within limitations.

Synonyms: CAS: 16423-68-0 ✲ FD&C RED NO. 3 ✲ SODIUM SALT OF IODEOSIN

ESCULIN

Products: Derived from leaves and bark of horse chestnut tree. In skin lotions, creams, and liquids.

Use: A skin protectant.

Health Effects: Could cause an allergic reaction in susceptible individuals. Harmless when used for intended purposes.

Synonyms: CAS: 531-75-9 ❊ ESCOSYL ❊ ESCULOSIDE ❊ 6,7-DIHYDROXYCOUMARIN ❊ 6-GLUCOSIDE

ESPARTO

Products: For carbon paper and other high quality papers. Used for paper money.

Use: Wax that is used as a substitute for carnauba wax. Derived from plant leaves of high cellulose content.

Health Effects: Could cause allergic reaction.

Synonyms: None found.

ESSENTIAL OIL

Products: In perfumes, air fresheners, sprays, deodorants, e.g. oil of wintergreen, balsams, and oil of bitter almond. Previously obtained from leaves, flowers, bark, or stems of plants. They are mostly synthetic now.

Use: As odorant.

Health Effects: Highly allergenic. Could cause eye or skin irritation upon contact. Possible respiratory effects when breathed.

Synonyms: See specific examples.

ESTERASE-LIPASE

Products: In cheese, fats, milk products, and oil.

Use: As a flavor enhancer.

Health Effects: FDA approves use at moderate levels to accomplish the intended effect.

Synonyms: *Mucor miehei*

ESTERS OF FATTY ACIDS

Products: In adhesives, cosmetics, leather products, lubricants, and textile finishes.

Use: Various.

Health Effects: Harmless when used for intended purposes.

Synonyms: METHYL ESTER OF FATTY ACID ❖ BUTYL ESTER OF FATTY ACID ❖ PROPYL ESTER OF FATTY ACID ❖ GLYCERYL ESTER OF FATTY ACID ❖ POLYETHYLENE GLYCOL ESTER OF FATTY ACID

ETHANE

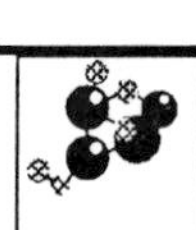

Products: Occurs in natural gas.

Use: As a refrigerant or fuel.

Health Effects: An asphyxiant gas. Severe fire risk if exposed to fire or open flame.

Synonyms: CAS: 74-84-0 ❖ DIMETHYL ❖ METHYLMETHANE

ETHANETHIOL

Products: As an LPG (liquified petroleum gas) odorant. In adhesives.

Use: Various. Deodorizer; tomato juice is reported to deodorize materials contaminated with this compound.

Health Effects: Toxic by swallowing and breathing. Flammable. Dangerous Fire risk. One of the most penetrating and persistent odors known (skunk).

Synonyms: CAS: 75-08-1 ❖ ETHYL SULFHYDRATE ❖ ETHYL MERCAPTAN

ETHANOLAMINE

Products: For detergents in dry cleaning, wool treatments, paints, polishes, and sprays.

Use: Various.

Health Effects: Moderately toxic by swallowing and skin contact. A corrosive irritant to skin eyes, nose, and throats. A mutagen (changes inherited characteristics).

Synonyms: CAS: 141-43-5 ❖ MEA ❖ MONOETHANOLAMINE ❖ COLAMINE ❖ 2-AMINOETHANOL ❖ 2-HYDROXYETHYLAMINE

ETHOHEXADIOL

Products: In insect repellent, printing inks, grooming, and hair care preparations.

Use: As an insecticide and for cosmetics.

Health Effects: Moderately toxic by swallowing and skin contact. A skin and severe eye irritant.

Synonyms: CAS: 94-96-2 ❖ 6-12 INSECT REPELLENT ❖ ETHYL HEXANEDIOL ❖ OCTYLENE GLYCOL

ETHOXYQUIN

Products: Sprayed on apples after picking for preservative. Slows spoiling of

Use: As a preservative and antioxidant (slows down spoiling due to

fruit. Prevents mold growth. | oxidation).

Health Effects: Toxic by swallowing. Wash fruits well before consuming!

Synonyms: CAS: 91-53-2 ✻ SANTOQUINE ✻ SANTOFLEX ✻ SANTOQUIN ✻ STOP-SCALD

2-ETHOXETHYL-p-METHOXYCINNAMATE

Products: In suntan preparations.

Use: As an ultraviolet (UV) absorber.

Health Effects: Could cause allergic reactions.

Synonyms: None found.

ETHYL ABIETATE

Products: In varnishes, lacquers, and coatings.

Use: Various.

Health Effects: An irritant.

Synonyms: None found.

ETHYL ACETATE

Products: In coffee (decaffeinated), fruits, tea (decaffeinated), and vegetables.

Use: As coloring agent, flavoring agent, solvent, synthetic flavoring substance, and additive.

Health Effects: Poison by breathing pure chemical. In large amounts it is mildly toxic by swallowing. Effects by breathing are: olfactory changes (ability of nose to smell), conjunctiva irritation (nose, throat, eyelid lining), and pulmonary changes (lungs and respiratory). A mutagen (changes inherited characteristics). Irritating to nose and throat surfaces, particularly the eyes, gums, and respiratory passages, and is also mildly narcotic. It can cause dermatitis. FDA permits use within stated limits.

Synonyms: CAS: 141-78-6 ✻ ACETIC ETHER ✻ ACETIDIN ✻ ACETOXYETHANE ✻ ETHYL ACETIC ESTER ✻ ETHYL ETHANOATE ✻ VINEGAR NAPHTHA

ETHYL ACRYLATE

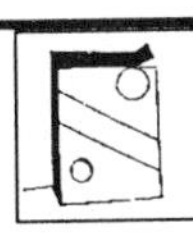

Products: As flavoring agent and for packaging materials.

Use: Various.

Health Effects: Confirmed carcinogen (causes cancer). In large amounts it is poison by swallowing and breathing. Moderately toxic by skin contact. Effects of breathing are: eye, olfactory (nose and smelling) and pulmonary (lung) changes. A skin and eye irritant. A substance which migrates to food from packaging materials. FDA approves use at moderate levels to accomplish the intended effect.

Synonyms: CAS: 140-88-5 ✲ ACRYLIC ACID ETHYL ESTER ✲ ETHOXYCARBONYLETHYLENE ✲ ETHYL PROPENOATE ✲ ETHYL-2-PROPENOATE

ETHYL ALCOHOL

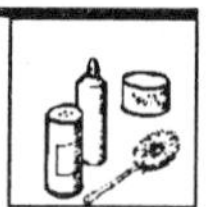

Products: In antimicrobial agent, detergents, cleaning preparations, cosmetics, gasohol, antifreeze, windshield defroster spray, and beverages.

Use: Various

Health Effects: It is a confirmed carcinogen (causes cancer) by drinking of beverage alcohol. Moderately toxic to humans by swallowing. Mildly toxic by breathing and skin contact. Swallowing effects include sleep disorders, hallucinations, distorted perceptions, convulsions, motor activity changes, ataxia, coma, antipsychotic, headache, pulmonary changes, alteration in gastric secretion, nausea or vomiting. Other effects include gastrointestinal changes, menstrual cycle changes and body temperature decrease. Can also cause glandular effects in humans. Reproductive effects (infertility, or sterility, or birth defects) by swallowing. The FAS (fetal alcohol syndrome) effects on newborns as a result of the mother drinking during pregnancy include: changes in Apgar score, neonatal measures or effects, and drug dependence. A mutagen (changes inherited characteristics). An eye and skin irritant.

Synonyms: CAS: 64-17-5 ✲ ABSOLUTE ETHANOL ✲ ALCOHOL ✲ ALCOHOL, anhydrous ✲ ALCOHOL, dehydrated ✲ ALGRAIN ✲ ANHYDROL ✲ COLOGNE SPIRIT ✲ ETHANOL 200 PROOF ✲ ETHYL HYDRATE ✲ ETHYL HYDROXIDE ✲ FERMENTATION ALCOHOL ✲ GRAIN ALCOHOL ✲ METHYLCARBINOL ✲ MOLASSES ALCOHOL ✲ POTATO ALCOHOL

ETHYLAMINE

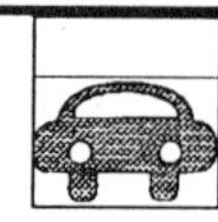

Products: In petroleum and detergents.

Use: Various.

Health Effects: A strong irritant. Flammable.

Synonyms: CAS: 75-04-7 ✲ MONOETHYLAMINE ✲ AMINOETHANE

ETHYL ANTHRANILATE

Products: In fruit and floral flavors and aroma for beverages, ices, creams, baked products, and gelatins.

Use: As a perfume agent or flavoring agent. An ingredient that affects the taste or smell of final product.

Health Effects: Concentrated chemical is moderately toxic by swallowing gross amounts. A skin irritant. FDA approves use at moderate levels to accomplish the intended effect.

Synonyms: CAS: 87-25-2 ✲ o-AMINOBENZOIC ACID, ETHYL ESTER ✲ ETHYL-o-AMINOBENZOATE

ETHYL BENZOATE

Products: In beverages, ice creams, ices, candies, bakery products, gum, gelatins, and liquors,

Use: As artificial fruit (strawberry, cherry, grape, raspberry) and nut flavors.

Health Effects: In large amounts pure substance is moderately toxic by swallowing. Mildly toxic by skin contact. A skin and eye irritant.

Synonyms: CAS: 93-89-0 ✲ BENZOIC ACID ETHYL ESTER ✲ ETHYL BENZENECARBOXYLATE ✲ BENZOIC ETHER ✲ ESSENCE OF NIOBE

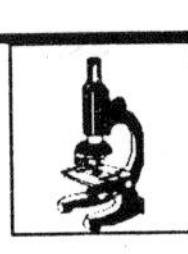

ETHYL BROMIDE

Products: A medical anesthetic, refrigerant, and solvent.

Use: Various.

Health Effects: Toxic by swallowing, breathing, and skin absorption. A strong irritant.

Synonyms: CAS: 74-96-4 ✲ BROMOETHANE

2-ETHYL BUTYL ALCOHOL

Products: For oils, waxes, and dyes.

Use: Perfumes and flavoring agent and as a solvent.

Health Effects: Harmless when used for intended purposes. Combustible. Moderate fire risk.

Synonyms: CAS: 97-95-0 ✲ 2-ETHYL BUTANOL ✲ PSEUDOHEXYL ALCOHOL

ETHYL BUTYRATE

Products: In fruit, caramel, nut, beverage eggnog flavorings, ices, candies, bakery products, and puddings.

Use: As flavoring and in perfumery. An ingredient to affect the taste or smell of final product.

Health Effects: In large amounts pure chemical is mildly toxic by swallowing. A skin irritant. FDA approves use at a moderate level.

Synonyms: CAS: 105-54-4 ✲ BUTANOIC ACID ETHYL ESTER ✲ BUTYRIC ETHER ✲ ETHYL BUTANOATE

ETHYL CAPROATE

Products: In artificial fruit essences.

Use: As a flavoring agent.

Health Effects: A skin irritant. FDA approves use in moderate amounts.

Synonyms: CAS: 123-66-0 ✲ ETHYL BUTYLACETATE ✲ ETHYL HEXANOATE

ETHYL CELLULOSE

Products: Confectionery, eggs (shell), food supplements in tablet form, gum, hot-glue adhesives, and printing inks.

Use: As a binder and filler in dry vitamin preparations. As a color fixer, fixative, in flavoring compounds; and a coating (protective component for vitamin and mineral tablets).

Health Effects: Harmless when used for intended purpose.

Synonyms: CAS: 9004-57-3 ✲ ETHYL ETHER OF CELLULOSE

ETHYL CINNAMATE

Products: In perfumes, spicy and oriental soaps, colognes, and bath products. Also in food spices.

Use: As a perfume fixative and food flavoring agent.

Health Effects: Moderately toxic by swallowing large amounts of pure substance. Could cause allergic effects.

Synonyms: CAS: 103-36-3 ✲ ETHYL-β -PHENYLACRYLATE ✲ ETHYL-3-PHENYLPROPENOATE ✲ CINNAMYLIC ETHER

ETHYL ENANTHATE

Products: In liqueurs and soft drinks.

Use: As cognac, berry, grape, cherry, apricot, and fruity-type soft drinks.

Health Effects: Harmless when used for intended purposes.

Synonyms: ETHYL HEPTANOATE ✲ COGNAC OIL ✲ ARTIFICIAL COGNAC ESSENCE ✲ OIL OF WINE

ETHYLENE DICHLORIDE

Products: In paint, varnish, gasoline, finish removers, soaps, scouring compounds, solvents, and fumigant.

Use: Various.

Health Effects: Toxic by swallowing, breathing, and skin absorption. Strong irritant to eyes and skin. A carcinogen (causes cancer).

Synonyms: CAS: 107-06-2 ✲ SYM-DICHLOROETHANE ✲ ETHYLENE CHLORIDE ✲ DUTCH OIL ✲ 1,2-DICHLOROETHANE

ETHYLENE GLYCOL

Products: For cosmetics (up to 5%), printing inks, wood stains, adhesives,

Use: Extremely varied including coolant, antifreeze, deicer, and moisturizer.

tobacco, airplanes, runways, ball point pen inks, stamp pad inks, and synthetic waxes.

Health Effects: Human poison by swallowing. Lethal dose for humans reported to be 100 ml. Effects by breathing and swallowing are: eye watering, anesthesia, headache, cough, nausea, vomiting, lung, kidney and liver changes. If swallowed it causes central nervous system stimulation followed by depression. Ultimately it causes potential lethal kidney damage. Very toxic in particulate form upon breathing. A mutagen (changes inherited characteristics). A skin, eye, nose, and throat irritant.

Synonyms: CAS: 107-21-1 ❊ ETHYLENE ALCOHOL ❊ GLYCOL ❊ 1,2-ETHANEDIOL

ETHYLENE GLYCOL METHYL ETHER

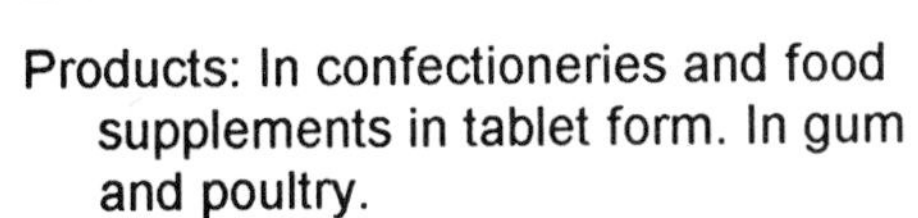

Products: In confectioneries and food supplements in tablet form. In gum and poultry.

Use: As a binder, color fixer, and extender.

Health Effects: Moderately toxic to humans by swallowing large amounts of pure substance. Effects by breathing are: change in motor activity, tremors, and convulsions. A skin and eye irritant. When used under conditions which do not require the application of heat, this material probably presents little hazard to health.

Synonyms: CAS: 109-86-4 ❊ GLYCOL ETHER EM ❊ GLYCOLMETHYL ETHER ❊ GLYCOL MONOMETHYL ETHER ❊ 2-METHOXYETHANOL ❊ METHOXYHYDROXYETHANE ❊ METHYL CELLOSOLVE ❊ METHYL ETHOXOL ❊ METHYL GLYCOL ❊ METHYL OXITOL

ETHYLENE GLYCOL DIACETATE

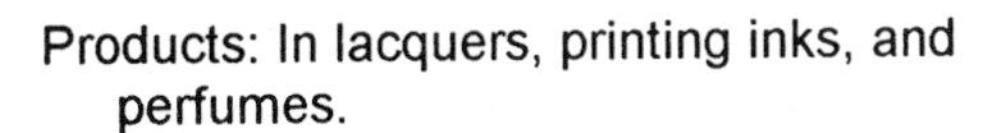

Products: In lacquers, printing inks, and perfumes.

Use: As a solvent and fixative.

Health Effects: Mildly toxic by swallowing and skin contact. An eye irritant.

Synonyms: CAS: 111-55-7 ❊ GLYCOL DIACETATE ❊ ETHLENE ACETATE ❊ ETHYLENE GLYCOL ACETATE

ETHYLENE GLYCOL MONOBUTYL ETHER

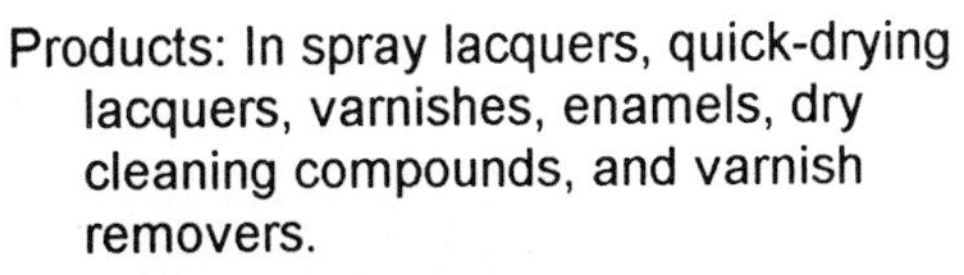

Products: In spray lacquers, quick-drying lacquers, varnishes, enamels, dry cleaning compounds, and varnish removers.

Use: As a solvent and in soap emulsifier.

Health Effects: Poison by swallowing and skin contact. Moderately toxic by breathing. Effects by breathing are; nausea, vomiting, headache, and nose tumors. A skin irritant.

Synonyms: CAS: 111-76-2 ✲ 2-BUTOXYETHANOL ✲ BUTYL CELLOSOLVE

ETHYLENE GLYCOL MONOETHYL ETHER

Products: In varnish remover, cleaning solutions, anti-icing additive for fuels.

Use: As a solvent and antifreeze.

Health Effects: Moderately toxic by swallowing and skin contact. Mildly toxic by breathing. An eye and skin irritant.

Synonyms: CAS: 110-80-5 ✲ 2-ETHOXYETHANOL ✲ CELLOSOVE SOLVENT

ETHYLENE GLYCOL MONOMETHYL ETHER

Products: In lacquers, enamels, varnishes, perfumes, wood stains, sealing moisture-proof cellophane, and jet-fuel deicing fluids.

Use: As a solvent, fixative, and additive.

Health Effects: Toxic by swallowing and breathing.

Synonyms: CAS: 109-86-4 ✲ 2-METHOXYMETHANOL ✲ METHYL CELLOSOLVE

ETHYLENEIMINE

Products: In adhesives, fuel oil, lubricants, and germicides.

Use: For microbial control, additive, and an intermediate (used to make other chemicals).

Health Effects: Corrosive. Absorbed by skin and causes tumors. A carcinogen (causes cancer). Could cause allergic skin reaction.

Synonyms: CAS: 151-56-4 ✲ AZRIDINE ✲ ETHYLENIMINE

ETHYL FORMATE

Products: Occurs naturally in apples and coffee extract. For lemonades, baked goods, candy (hard), candy (soft), chewing gum, fillings, frozen dairy desserts, gelatins, puddings, raisins, and Zante currents. In fumigants and larvicides.

Use: As a flavoring agent, fruit essence, and insecticide.

Health Effects: In concentration it is moderately toxic by swallowing large amounts. Mildly toxic by skin contact and breathing. A powerful breathing irritant in humans. A skin and eye irritant. Questionable carcinogen (may cause cancer). Highly flammable liquid. FDA approves use within limits.

Synonyms: CAS: 109-94-4 ❊ AREGINAL ❊ ETHYL FORMIC ESTER ❊ ETHYL METHANOATE ❊ FORMIC ACID, ETHYL ESTER ❊ FORMIC ETHER

2-ETHYLHEXYL ALCOHOL

Products: In paints, lacquers, mercerizing cotton, textile finishes, textile compounds, inks, and dry cleaning.

Use: Various; including penetrant, lubricant and solvent.

Health Effects: Moderately toxic by swallowing and skin contact. A severe eye and moderate skin irritant.

Synonyms: CAS: 104-76-7 ❊ 2-ETHYLHEXANOL ❊ OCTYLALCOHOL

2-ETHYLHEXYLAMINE

Products: In detergents, oil additives, and insecticides.

Use: Various.

Health Effects: Moderately toxic by swallowing, breathing, and skin contact. A severe skin and eye irritant.

Synonyms: CAS: 104-75-6

ETHYL HYDROXYETHYL CELLULOSE

Products: In protective coatings, gravure printing inks, and silk screen film.

Use: As a stabilizer, thickener, and binder.

Health Effects: Harmless when used for intended purposes.

Synonyms: EHEC ❊ CELLULOSE ETHER

ETHYL ISOBUTYRATE

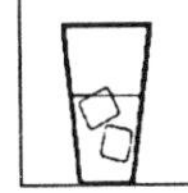

Products: In baked goods, beverages, and candy.

Use: As a flavoring agent.

Health Effects: A skin irritant. FDA approves use at moderate levels.

Synonyms: CAS: 97-62-1 ❊ ETHYL ISOBUTANOATE ❊ ETHYLISOBUTYRATE ❊ ETHYL-2-METHYLPROPANOATE ❊ ETHYL-2-METHYLPROPIONATE ❊ ISOBUTYRIC ACID, ETHYL ESTER ❊ 2-METHYLPROPIONIC ACID, ETHYL ESTER

ETHYL LAURATE

Products: In fruit, nut, liquor, spice, cheese flavorings, beverage flavorings, ices, creams, bakery products,

Use: As solvent and for food flavoring.

and gum.

Health Effects: Harmless when used for intended purposes.

Synonyms: CAS: 106-33-2 ✲ ETHYL DODECANOATE

ETHYL MALTOL

Products: In chocolates, desserts, and wines.

Use: As a flavoring agent and processing aid.

Health Effects: In gross amounts it is moderately toxic by swallowing. FDA approves use at moderate levels to accomplish the intended effects.

Synonyms: CAS: 4940-11-8 ✲ 2-ETHYL-3-HYDROXY-4H-PYRAN-4-ONE ✲ 2-ETHYL PYROMECONIC ACID ✲ 3-HYDROXY-2-ETHYL-4-PYRONE

ETHYLMERCURIC PHOSPHATE

Products: For seeds and wood.

Use: As a fungicide and preservative.

Health Effects: Poison by swallowing.

Synonyms: CAS: 2440-24-1 ✲ LIGNASAN FUNGICIDE ✲ CERESAN ✲ GRANOSAN

ETHYL METHYLPHENYLGLYCIDATE

Products: In beverages, candy, and ice cream.

Use: As a flavoring agent.

Health Effects: Mildly toxic by swallowing large amounts of pure substance. FDA approves use at moderate levels to accomplish the desired results.

Synonyms: CAS: 77-83-8 ✲ C-16 ALDEHYDE ✲ EMPG ✲ α-β-EPOXY-β-METHYLHYDROCINNAMIC ACID, ETHYL ESTER ✲ ETHYL α,β-EPOXY-β-METHYLHYDROCINNAMATE ✲ ETHYL 2,3-EPOXY-3-METHYL-3-PHENYLPROPIONATE ✲ ETHYL ESTER of 2,3-EPOXY-3-PHENYLBUTANOIC ACID ✲ FRAESEOL ✲ 3-METHYL-3-PHENYLGLYCIDIC ACID ETHYL ESTER ✲ STRAWBERRY ALDEHYDE

ETHYL NONANOATE

Products: In fruit and rum flavored beverages, beverages (alcoholic), candy, ice cream, bakery products, gelatins, puddings, desserts, and icings.

Use: As a synthetic flavoring agent.

Health Effects: Mildly toxic by swallowing gross amounts of concentrated chemical. A skin irritant. FDA approves use at moderate levels to accomplish the intended effect.

Synonyms: CAS: 123-29-5 ✲ ETHYL NONYLATE ✲ ETHYL PELARGONATE ✲ NONANOIC ACID, ETHYL ESTER ✲ WINE ETHER

ETHYL OXYHYDRATE

Products: In beverages, candy, ice cream, bakery goods, puddings, and gelatins.

Use: As a liquor, rum, or butter flavoring agent.

Health Effects: FDA approves use at moderate levels to accomplish the intended effect.

Synonyms: RUM ETHER ❊ SALICYLALDEHYDE

ETHYL PELARGONATE

Products: In alcoholic beverages (cognac essence), perfumes, and flame retardants.

Use: For flavorings and perfumery.

Health Effects: An irritant to eyes and lungs.

Synonyms: CAS: 2524-04-1 ❊ ETHYL NONANOATE ❊ WINE-ETHER ❊ ETHYL PCT

ETHYL PHENYLACETATE

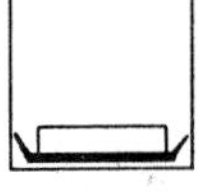

Products: In honey, butter, fruit flavoring, beverage flavoring, creams, ices, candy, and bakery products.

Use: For food product flavoring, perfume, aromas, and scents.

Health Effects: In large amounts it is moderately toxic by swallowing pure substance. FDA approves use at moderate levels to accomplish the intended effect.

Synonyms: CAS: 101-97-3 ❊ BENZENEACETIC ACID, ETHYL ESTER ❊ ETHYL BENZENEACETATE ❊ ETHYL PHENACETATE ❊ ETHYL-2-PHENYLETHANOATE ❊ ETHYL-α-TOLUATE ❊ PHENYLACETIC ACID, ETHYL ESTER ❊ α-TOLUIC ACID, ETHYL ESTER

ETHYLPHOSPHORIC ACID

Products: For metal items.

Use: As rust remover.

Health Effects: An irritant. Follow directions carefully when using products that contain this chemical.

Synonyms: None found.

ETHYL PROPIONATE

Products: In fruity, rum-scented candies, bakery products, and beverages.

Use: For flavoring agent and fruit syrups.

Health Effects: Moderately toxic by swallowing large amounts of concentrated chemical. A skin irritant. Flammable, a dangerous fire risk.

Synonyms: CAS: 105-37-3 ✲ PROPIONIC ETHER ✲ PROPIONIC ACID, ETHYLESTER

ETHYL SILICATE

Products: In coatings, paints, lacquers, and bondings.

Use: For weatherproof mortar, cements, bricks, and heat resistant paints.

Health Effects: Strong irritant to eyes, nose, and throat.

Synonyms: CAS: 78-10-4 ✲ TETRAETHYL ✲ ORTHOSILICATE

ETHYL VANILLIN

Products: In baked goods, beverages, ice cream, and sauces.

Use: For flavoring agent and perfume ingredient.

Health Effects: Pure substance is moderately toxic by swallowing large amounts. A skin irritant. FDA approves use at a moderate level to accomplish the desired results.

Synonyms: CAS: 121-32-4 ✲ BOURBONAL ✲ ETHAVAN ✲ ETHOVAN ✲ 3-ETHOXY-4-HYDROXYBENZALDEHYDE ✲ ETHYLPROTAL ✲ 4-HYDROXY-3-ETHOXYBENZALDEHYDE ✲ PROTOCATECHUIC ALDEHYDE ETHYL ETHER ✲ QUANTROVANIL ✲ VANILLAL ✲ VANIROM

EUCALYPTUS OIL

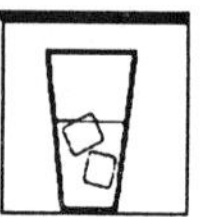

Products: In bakery products, beverages (alcoholic), beverages (nonalcoholic), confections, and ice cream. In pharmaceuticals and antibacterials.

Use: For flavoring agent and odorant. An ingredient that affects the taste or smell of final product.

Health Effects: A poison by swallowing large amounts of concentrated oil. Effects by swallowing are: eye spasms, sleepiness, and respiratory depression. A skin irritant. FDA approves use at moderate levels to accomplish the desired results.

Synonyms: CAS: 8000-48-4 ✲ DINKUM OIL ✲ EUKALYPTUS OEL ✲ OIL of EUCALYPTUS

EUGENOL

Products: In baked goods, beverages (nonalcoholic), candy, chewing gum, condiments, confections, gelatin desserts, ice cream, meat, puddings, and spice oils. In medications.

Use: As a flavoring agent and in perfumery to replace oil of cloves. An insect attractant. As dental analgesic (pain killer).

Health Effects: In large amounts it is moderately toxic by swallowing pure substance. A mutagen (changes inherited characteristics). A skin irritant. GRAS (generally regarded as safe).

Synonyms: CAS: 97-53-0 ✲ 4-ALLYLGUAIACOL ✲ 4-ALLYL-1-HYDROXY-2-METHOXYBENZENE ✲ 4-ALLYL-2-METHOXYPHENOL ✲ CARYOPHYLLIC ACID ✲ EUGENIC ACID ✲ SYNTHETIC EUGENOL

"Food allergy symptoms are stomach pain, diarrhea, nausea, vomiting, cramps, hives, rash, and occasionally hay-fever type reactions."

The Columbia University College of Physicians & Surgeons Complete Home Medical Guide

FARNESOL

Products: In beverages, ice creams, ices, bakery products, and puddings. Occurs naturally in anise, flowers, oils, and roses.

Use: As apricot, banana, melon and berry flavoring.

Health Effects: FDA approves use in limited amounts. Mildly toxic by swallowing gross amounts of pure substance. A mutagen (changes inherited characteristics)

Synonyms: CAS: 4602-84-0 ✲ FARNESYL ALCOHOL ✲ 3,7,11-TRIMETHYL-2,6,10-DODECATRIEN-1-OL

FATTY ACIDS

Products: In soaps, detergents, bubble baths, lipstick, candles, salad oils, shortenings, and cosmetics.

Use: Component in the manufacture of food-grade additives, defoaming agent, lubricant, emulsifier, and binder.

Health Effects: Harmless when used for intended purpose.

Synonyms: CAPRIC ACID ✲ CAPRYLIC ACID ✲ LAURIC ACID ✲ MYRISTIC ACID ✲ OLEIC ACID ✲ PALMITIC ACID ✲ STEARIC ACID

FD&C BLUE No. 1

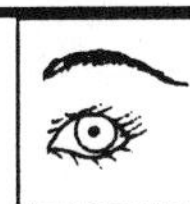

Products: In baked goods, candy, confections, food, drugs, and cosmetics.

Use: Color additive.

Health Effects: A possible carcinogen (causes cancer) in concentration. Could cause allergic reaction. FDA approves use at moderate levels to accomplish the desired results.

Synonyms: CAS: 38444-45-9 ✲ ACID SKY BLUE A ✲ AIZEN FOOD BLUE No. 2 ✲ BRILLIANT BLUE FCD No. 1 ✲ BRILLIANT BLUE FCF ✲ CANACERT BRILLIANT BLUE FCF ✲ COSMETIC BLUE LAKE ✲ D&C BLUE No. 4 ✲

DOLKWAL BRILLIANT BLUE ❖ FOOD BLUE 2 ❖ FOOD BLUE DYE No. 1 ❖ HEXACOL BRILLIANT BLUE A ❖ INTRACID PURE BLUE L

FD&C BLUE No. 2

Products: In beverages, baked goods, cereals, confections, jellies, and candies.

Use: As a color additive.

Health Effects: Moderately toxic by swallowing gross amounts of pure substance. Could cause allergic reaction.

Synonyms: CAS: 860-22-0 ❖ ACID BLUE W ❖ ACID LEATHER BLUE IC ❖ A.F. BLUE No. 2 ❖ AIRDALE BLUE IN ❖ AMACID BRILLIANT BLUE ❖ ANILINE CARMINE POWDER ❖ DISODIUM INDIGO-5,5-DISULFONATE ❖ INDIGO CARMINE ❖ INDIGO EXTRACT ❖ MAPLE INDIGO CARMINE ❖ SOLUBLE INDIGO

FD&C GREEN No. 3

Products: In beverages, cereals, desserts, soft drinks, food, drugs, and cosmetics.

Use: As a color additive.

Health Effects: Approved for use at moderate levels in food, drugs, and cosmetics, except in the eye area.

Synonyms: CAS: 2353-45-9 ❖ AIZEN FOOD GREEN No. 3 ❖ C.I. FOOD GREEN 3 ❖ FAST GREEN FCF ❖ SOLID GREEN FCF

FD&C RED No. 3

Products: In candy, cherries, confections, toothpaste, cereals, gelatins, and printing inks.

Use: As a color additive.

Health Effects: In gross amounts it is moderately toxic by swallowing pure substance . A possible carcinogen (may cause cancer). A human mutagen (changes inherited characteristics).

Synonyms: CAS: 16423-68-0 ❖ AIZEN ERYTHROSINE ❖ CALCOCID ERYTHROSINE N ❖ CANACERT ERYTHROSINE BS ❖ 9-(o-CARBOXYPHENYL)-6-HYDROXY-2,4,5,7-TETRAIODO-3-ISOXANTHONE ❖ CILEFA PINK B ❖ D&C RED No. 3 ❖ DOLKWAL ERYTHROSINE ❖ DYE FD&C RED No. 3 ❖ ERYTHROSIN ❖ MAPLE ERYTHROSINE ❖ 2′4′,5′,7′-TETRAIODOFLUORESCEIN, DISODIUM SALT ❖ TETRAIODOFLUORESCEIN SODIUM SALT ❖ USACERT RED No. 3

FD&C RED No. 40

Products: In beverages, candy, cereals, confections, desserts, pharmaceuticals, cosmetics, and ice cream.

Use: As a color additive.

Health Effects: Approved for use at moderate levels to accomplish the desired results.

Synonyms: CAS: 25956-17-6 ✼ ALLURA RED AC ✼ C.I. 16035

FD&C YELLOW No. 3

Products: In beverages, cereals, baked goods, puddings, hair coloring, and permanent hair products.

Use: As a color additive in food and dyes.

Health Effects: Moderately toxic by swallowing gross amounts of pure chemical. Possible carcinogen (may cause cancer). Could cause allergic reaction.

Synonyms: CAS: 85-84-7 ✼ A.F. YELLOW No. 2 ✼ 1-BENZENE-AZO-β-NAPHTHYLAMINE ✼ 1-BENZENEAZO-2-NAPHTHYLAMINE ✼ CERISOL YELLOW AB ✼ DOLKWAL YELLOW AB ✼ EXT. D&C YELLOW No. 9 ✼ GRASAL YELLOW ✼ JAUNE AB ✼ OIL YELLOW A ✼ 1-(PHENYLAZO)-2-NAPHTHALENAMINE ✼ 1-(PHENYLAZO)-2-NAPHTHYLAMINE ✼ YELLOW AB ✼ YELLOW No. 2

FD&C YELLOW No. 5

Products: In butter, cheese, ice cream, and ink

Use: A color additive.

Health Effects: Suspected to be the cause of allergy, especially in aspirin-sensitive individuals. In gross amounts it is mildly toxic by swallowing concentrated substance. Effects on the body by swallowing excessive quantities are paresthesia (abnormal sensation of burning and tingling) and changes in teeth and supporting structures. A human mutagen (changes inherited characteristics).

Synonyms: CAS: 1934-21-0 ✼ ACID LEATHER YELLOW T ✼ AIREDALE YELLOW T ✼ AIZEN TARTRAZINE ✼ ATUL TARTRAZINE ✼ BUCACID TARTRAZINE ✼ D&C YELLOW No. 5 ✼ DOLKWAL TARTRAZINE ✼ EGG YELLOW A ✼ EUROCERT TARTRAZINE ✼ FOOD YELLOW No. 4 ✼ HEXACOL TARTRAZINE ✼ HYDRAZINE YELLOW ✼ KARO TARTRAZINE ✼ LAKE YELLOW ✼ MAPLE TARTRAZOL YELLOW ✼ TARTAR YELLOW ✼ TARTRAZINE ✼ TARTRAZOL YELLOW ✼ TRISODIUM-3-CARBOXY-5-HYDROXY-1-p-SULFOPHENYL-4-p-SULFOPHENYLAZOPYRAZOLE ✼ VONDACID TARTRAZINE ✼ WOOL YELLOW ✼ YELLOW LAKE 69

FD&C YELLOW No. 6

Products: In bakery goods, beverages, butter, candy, cereals, cheese, confections, desserts, and ice cream.

Use: A color additive in food.

Health Effects: A possible carcinogen (may cause cancer) in gross amounts. Could cause allergic reaction. FDA approves use at moderate levels to accomplish the desired results.

Synonyms: CAS: 2783-94-0 ✼ ACID YELLOW TRA ✼ AIZEN FOOD YELLOW No. 5 ✼ CANACERT SUNSET YELLOW FCF ✼ GELBORANGE-S (GERMAN) ✼ SUNSET YELLOW FCF

FELDSPAR

Products: In soaps, cements, tarred roofing materials, pottery, enamelware, ceramic ware, glass, and fertilizer.

Use: As a filler, grit, and glazing material.

Health Effects: Toxic as fine-ground powder.

Synonyms: POTASSIUM ALUMINOSILICATE

FENNEL OIL

Products: In seasoning for sausage, beverages, fruit products, bakery products, meats, root beer, and condiments.

Use: As a food and beverage flavoring or seasoning.

Health Effects: Moderately toxic by swallowing gross amounts. A mutagen (changes inherited characteristics). A severe skin irritant. Could cause allergic reaction. GRAS (generally regarded as safe) when used at moderate levels to accomplish the desired results.

Synonyms: CAS: 8006-84-6 ✲ BITTER FENNEL OIL ✲ FENCHEL OEL (GERMAN) ✲ OIL of FENNEL

FENURON

Products: A weed and brush killer.

Use: As a herbicide.

Health Effects: Moderately toxic by swallowing. A mutagen (changes inherited characteristics).

Synonyms: CAS: 101-42-8 ✲ FENULON ✲ DIBAR ✲ PUD ✲ FENIDIN ✲ 1,1-DIMETHYL-3-PHENYLUREA

FERRIC ACETATE, BASIC

Products: For wood, leather, and medicine.

Use: In preservatives, inks, and dye.

Health Effects: It is important to follow directions carefully on products that contain this chemical.

Synonyms: CAS: 10450-55-2 ✲ IRON ACETATE, BASIC

FERRIC AMMONIUM CITRATE

Products: In medications, photography,

Use: As an anticaking agent, dietary

and feed additive.

supplement, and nutrient.

Health Effects: Harmless when used for intended purposes and within stated limits.

Synonyms: CAS: 1333-00-2 ❃ FERRIC AMMONIUM CITRATE, GREEN ❃ IRON(III) AMMONIUM CITRATE

FERRIC CHLORIDE, ANHYDROUS

Products: In pigments, inks, astringents, styptics, dyes, and inks.

Use: As disinfectant, sewage deodorizer, and water purifier.

Health Effects: Toxic by swallowing. A strong irritant to skin and tissue.

Synonyms: CAS: 7705-08-0 ❃ FERRIC TRICHLORIDE ❃ FERRIC PERCHLORIDE ❃ IRON CHLORIDE ❃ IRON TRICHLORIDE ❃ IRON PERCHLORIDE

FERRIC PHOSPHATE

Products: In egg substitutes (frozen), pasta products, and rice products.

Use: As a dietary supplement and nutrient supplement.

Health Effects: GRAS (generally regarded as safe).

Synonyms: CAS: 10045-86-0 ❃ FERRIC ORTHOPHOSPHATE ❃ IRON PHOSPHATE

FERRIC STEARATE

Products: In varnishes and photocopy machine chemicals.

Use: As a drier.

Health Effects: Could cause allergic reaction to skin. A possible skin irritant.

Synonyms: IRON STEARATE

FERRIC SULFATE

Products: In personal care, hair coloring, and dye products,

Use: As a deodorant, astringent, disinfectant, and textile printing inks.

Health Effects: GRAS (generally regarded as safe).

Synonyms: CAS: 10028-22-5 ❃ DIIRON TRISULFATE ❃ IRON PERSULFATE ❃ IRON SESQUISULFATE ❃ IRON SULFATE (2:3) ❃ IRON(III) SULFATE ❃ IRON TERSULFATE ❃ SULFURIC ACID, IRON (3^{+}) SALT (3:2)

FERROUS FUMARATE

Products: In cereals and waffles (frozen).

Use: As a dietary supplement and nutritional supplement.

Health Effects: In gross amounts it is moderately toxic by swallowing pure chemical. GRAS (generally regarded as safe). Limited as a source of iron for special dietary purposes.

Synonyms: CAS: 141-01-5 ❖ CPIRON ❖ ERCO-FER ❖ ERCOFERRO ❖ FEOSTAT ❖ FEROTON ❖ FERROFUME ❖ FERRONAT ❖ FERRONE ❖ FERROTEMP ❖ FERRUM ❖ FERSAMAL ❖ FIRON ❖ FUMAFER ❖ FUMAR-F ❖ FUMIRON ❖ GALFER ❖ HEMOTON ❖ IRCON ❖ IRON FUMARATE ❖ METERFER ❖ METERFOLIC ❖ ONE-IRON ❖ PALAFER ❖ TOLERON ❖ TOLFERAIN ❖ TOLIFER

FERROUS GLUCONATE

Products: For olives (ripe) and vitamin pills.

Use: As a color additive, dietary supplement, and nutrient supplement.

Health Effects: Moderately toxic by swallowing gross amounts of concentrated chemical. Effects on the body by swallowing are diarrhea, nausea, and vomiting. GRAS (generally regarded as safe).

Synonyms: CAS: 299-29-6 ❖ FERGON ❖ FERGON PREPARATIONS ❖ FERLUCON ❖ FERRONICUM ❖ GLUCO-FERRUM ❖ IROMIN ❖ IRON GLUCONATE ❖ IROX (GADOR) ❖ NIONATE ❖ RAY-GLUCIRON

FERROUS OXALATE

Products: In pigments and photographic developer.

Use: For coloration.

Health Effects: Toxic.

Synonyms: CAS: 516-03-0 ❖ IRON OXALATE

FERROUS SULFATE

Products: In wood preservatives and in wine.

Use: As a food additive, dietary supplement, nutrient, among others.

Health Effects: It is a human poison by swallowing large amounts. Effects on the body by swallowing are aggression, somnolence, brain recording changes, diarrhea, nausea, vomiting, bleeding from the stomach, and coma. A mutagen (changes inherited characteristics). GRAS (generally regarded as safe) when used at moderate levels.

Synonyms: CAS: 7720-78-7 ❖ COPPERAS ❖ DURETTER ❖ DUROFERON ❖ EXSICCATED FERROUS SULPHATE ❖ FEOSOL ❖ FEOSPAN ❖ FER-IN-SOL ❖ FERRO-GRADUMET ❖ FERROSULFATE ❖ FERRO-THERON ❖ FERSOLATE ❖ GREEN VITRIOL ❖ IRON MONOSULFATE ❖ IRON PROTOSULFATE ❖ IRON(II) SULFATE (1:1) ❖ IRON VITRIOL ❖ IROSPAN ❖ IROSUL ❖ SLOW-FE ❖ SULFERROUS ❖ SULFURIC ACID, IRON(2^+) SALT (1:1)

FERROUS SULFATE HEPTAHYDRATE

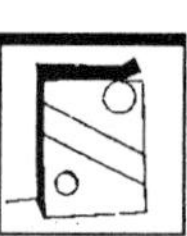

Products: In baking mixes, cereals, infant foods, pasta products, and wine.

Use: As a clarifying agent, dietary supplement, nutrient supplement, processing aid, and stabilizer.

Health Effects: Moderately toxic by swallowing pure, gross amounts. A mutagen (changes inherited characteristics). GRAS (generally regarded as safe) when used within limitations (in wine).

Synonyms: CAS: 7782-63-0 ❋ COPPERAS ❋ FEOSOL ❋ FER-IN-SOL ❋ FERO-GRADUMET ❋ FERROUS SULFATE (FCC) ❋ FESOFOR ❋ FESOTYME ❋ GREEN VITROL ❋ HAEMOFORT ❋ IRONATE ❋ IRON(II) SULFATE (1:1), HEPTAHYDRATE ❋ IRON VITROL ❋ IROSUL ❋ PRESFERSUL ❋ SULFERROUS

FICIN

Products: Derived from the fig tree *Ficus*. In beer, cereals (precooked), meat (raw cuts), poultry, and wine.

Use: For chillproofing of beer, coagulation of milk, enzyme, meat tenderizing, processing aid, and tenderizing agent.

Health Effects: Mildly toxic by swallowing pure, gross amounts.

Synonyms: CAS: 9001-33-6 ❋ DEBRICIN ❋ FICUS PROTEASE ❋ FICUS PROTEINASE ❋ HIGUEROXYL DELABARRE ❋ TL 367

FIR NEEDLE OIL, CANADIAN TYPE

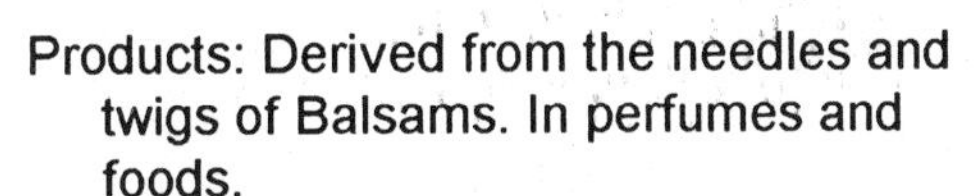

Products: Derived from the needles and twigs of Balsams. In perfumes and foods.

Use: As a fragrance, odorant, and flavoring.

Health Effects: FDA approves use at moderate levels to accomplish the desired results. Could cause allergic reactions.

Synonyms: BALSAM FIR OIL

FIR NEEDLE OIL, SIBERIAN

Products: Found in the needles and twigs of pine trees. In perfumes and foods.

Use: As a fragrance, odorant, and flavoring.

Health Effects: Mildly toxic by swallowing. A skin irritant. FDA approves use at moderate levels to accomplish the intended effect. Could cause allergic reactions.

Synonyms: CAS: 8021-29-2 ❋ PINE NEEDLE OIL

FISH GLUE

Products: For gummed tape, cartons, blueprint paper, and printing plates. A ton of skins yields about 50 gallons of liquid glue.

Use: An adhesive.

Health Effects: Harmless when used for intended purposes. Susceptible individuals have experienced skin irritation.

Synonyms: None found.

FLUORIDE

Products: In toothpastes, teeth rinses, and dentifrices.

Use: Tooth cavity preventitive.

Health Effects: Harmless when used at rate of 1 ppm in drinking water for the purpose of reducing tooth decay. Excessive doses can cause nausea, vomiting, diarrhea, and cramps. Can aggravate asthma and cause severe bone changes making movement painful. Irritant to the eyes, skin, nose, and throats.

Synonyms: FLUOSILICIC ACID ✲ SODIUM SILICOFLUORIDE ✲ SODIUM FLUORIDE

FLUORSPAR

Products: In cements, dentifrices, sanding wheels, paint pigments, and wood preservatives.

Use: As a filler, abrasive, and catalyst.

Health Effects: Harmless when used for appropriate purposes.

Synonyms: FLUORITE ✲ FLORSPAR

FLUOSILICIC ACID

Products: In cement, wood preservative, and disinfectant.

Use: For hardening and sterilizing.

Health Effects: Extremely corrosive by skin contact and breathing.

Synonyms: CAS: 16961-83-4 ✲ HYDROFLUOSILICIC ACID ✲ FLUOROSILIC ACID ✲ HEXAFLUORISILIC ACID ✲ HYDROGEN HEXAFLUORISILICATE

FOLIC ACID

Products: Found in green vegetables, liver, kidney, and nuts. Added to cosmetics.

Use: As a dietary supplement, nutrient, skin softener.

Health Effects: A member of the vitamin B complex. Very important during early pregnancy. FDA approves use within limitations.

Synonyms: CAS: 59-30-3 ✲ l-N-(p-(((-2-AMINO-4-HYDROXY-6-PTERIDINYL)METHYL)AMINO)BENZOYL)GLUTAMIC ACID ✲ FOLACIN ✲ FOLATE ✲ FOLCYSTEINE ✲ PTEGLU ✲ PTEROYLGLUTAMIC ACID ✲ PTEROYL-l-GLUTAMIC ACID ✲ PTEROYLMONOGLUTAMIC ACID ✲ PTEROYL-l-MONOGLUTAMIC ACID ✲ VITAMIN Bc ✲ VITAMIN M

FOLPET

Products: In paints and enamels.

Use: As a fungicide and bactericide.

Health Effects: Moderately toxic by swallowing. A mutagen (changes inherited characteristics).

Synonyms: CAS: 133-07-3 ✲ PHALTAN ✲ FOLPAN ✲ THIOPHAL ✲ ORTHOPHALTAN

FOOD STARCH, MODIFIED

Products: In desserts, gravies, pie fillings, and sauces.

Use: As a binder, glue-like stabilizer, and thickening agent.

Health Effects: Harmless when used for intended purposes.

Synonyms: None found.

FORMALDEHYDE

Products: In glues, air fresheners, antiperspirants, dry cleaning solvents, fingernail polish, hair spray, laundry spray starch, perfumes, after shave lotions, preservatives, cottonseed, packaging materials, fabric durable press treatment, particle board, and plywood.

Use: As a deodorizer, solvent, preservative, disinfectant, and adhesive among others.

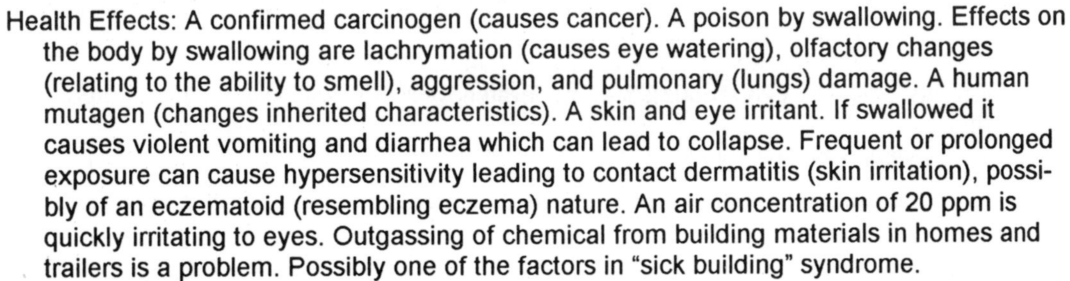

Health Effects: A confirmed carcinogen (causes cancer). A poison by swallowing. Effects on the body by swallowing are lachrymation (causes eye watering), olfactory changes (relating to the ability to smell), aggression, and pulmonary (lungs) damage. A human mutagen (changes inherited characteristics). A skin and eye irritant. If swallowed it causes violent vomiting and diarrhea which can lead to collapse. Frequent or prolonged exposure can cause hypersensitivity leading to contact dermatitis (skin irritation), possibly of an eczematoid (resembling eczema) nature. An air concentration of 20 ppm is quickly irritating to eyes. Outgassing of chemical from building materials in homes and trailers is a problem. Possibly one of the factors in "sick building" syndrome.

Synonyms: CAS: 50-00-0 ✲ BFV ✲ FA ✲ FANNOFORM ✲ FORMALDEHYDE, solution (DOT) ✲ FORMALIN ✲ FORMALIN 40 ✲ FORMALIN (DOT) ✲ FORMALITH ✲ FORMIC ALDEHYDE ✲ FORMOL ✲ FYDE ✲ HOCH ✲ IVALON ✲ KARSAN ✲ LYSOFORM ✲ METHANAL ✲ METHYL ALDEHYDE ✲ METHYLENE GLYCOL ✲ METHYLENE OXIDE ✲ MORBOCID ✲ OXOMETHANE ✲ OXYMETHYLENE ✲ PARAFORM ✲ POLYOXYMETHYLENE GLYCOLS ✲ SUPERLYSOFORM

FORMIC ACID

Products: In food packaging, synthetic

Use: As a fumigant, brewing antiseptic,

flavoring, hair products, perfumes, lacquers, and refrigerant.

solvent, and plasticizer.

Health Effects: Moderately toxic by swallowing. Mildly toxic by breathing. A mutagen (changes inherited characteristics). Corrosive. A skin and severe eye irritant. A substance migrating to food from packaging materials. GRAS (generally regarded as safe) as an indirect additive.

Synonyms: CAS: 64-18-6 ✲ AMINIC ACID ✲ FORMYLIC ACID ✲ HYDROGEN CARBOXYLIC ACID ✲ METHANOIC ACID

FUMARIC ACID

Products: In printing inks, beverage mixes (dry), candy, desserts, pie fillings, poultry, and wine.

Use: As an acidifier, tartness agent, curing accelerator, and flavoring agent.

Health Effects: In large amounts it is mildly toxic by swallowing pure substance. Toxic by skin contact. A skin and eye irritant. FDA approves use at moderate levels to accomplish the intended effect when used within limits..

Synonyms: CAS: 110-17-8 ✲ ALLOMALEIC ACID ✲ BOLETIC ACID ✲ trans-BUTENEDIOIC ACID ✲ (E)-BUTENEDIOIC ACID ✲ trans-1,2-ETHYLENEDICARBOXYLIC ACID ✲ (E)1,2-ETHYLENEDICARBOXYLIC ACID ✲ LICHENIC ACID

FURCELLERAN GUM

Products: In flans, jams, jellies, meat products (gelled), puddings (milk), and toothpastes. Harmless when used for intended purposes.

Use: As an emulsifier, stabilizer, thickening agent, and gelling agent.

Health Effects: In excessive amounts it is moderately toxic by swallowing concentrated chemical.

Synonyms: CAS: 9000-21-9 ✲ BURTONITE 44

FURFURAL

Products: In shoe dyes, leather coloring, weed killer, and fungicide. Also in caramel, butterscotch, fruit, and molasses flavors. In nut flavors for beverages and ice cream dessert syrups.

Use: As herbicide, leather preservative, and flavoring.

Health Effects: Poison by swallowing large amounts. Moderately toxic by breathing and skin contact. A human mutagen (changes inherited characteristics). A skin and eye irritant. The liquid is dangerous to the eyes. The vapor is irritating to nose and throats and is a central nervous system poison. GRAS (generally regarded as safe) when used in moderate amounts.

Synonyms: CAS: 98-01-1 ✻ ARTIFICIAL ANT OIL ✻ FURAL ✻ FURALE ✻ 2-FURALDEHYDE ✻ 2-FURANALDEHYDE ✻ 2-FURANCARBONAL ✻ 2-FURANCARBOXALDEHYDE ✻ 2-FURFURAL ✻ FURFURALDEHYDE ✻ FURFUROL ✻ FURFUROLE ✻ FUROLE ✻ α-FUROLE ✻ 2-FURYL-METHANAL ✻ PYROMUCIC ALDEHYDE

FUSEL OIL

Products: In explosives; varnishes, lacquers, fats, oils, waxes, and perfumery. In liquor, wine, whiskey, food, and beverage seasoning.

Use: As solvent, flavoring, and gelatinizing agent.

Health Effects: May contain carcinogens. Toxic by swallowing and breathing. FDA states use should be at moderate level to accomplish the desired results.

Synonyms: CAS: 8013-75-0 ✻ FUSEL OIL, REFINED

G

"Women who dye their hair increase their risk of lymphoma by 50 percent."
American Journal of Public Health/National Cancer Institute

GALLIC ACID

Products: In inks, dyes, tanning agents, photography chemicals, and pharmaceuticals.

Use: As marker colors, leather coloring, and dyeing.

Health Effects: Mildly toxic by swallowing. Could cause allergic reaction.

Synonyms: CAS: 149-91-7 ✲ 3,4,5-TRIHYDROXYBENZOIC ACID

GARLIC OIL

Products: In meat, sauces, and vegetables.

Use: As a flavoring agent.

Health Effects: An eye irritant. GRAS (generally regarded as safe) when used in moderate levels.

Synonyms: None found.

GERANIUM OIL ALGERIAN TYPE

Products: In bakery products, beverages (nonalcoholic), chewing gum, confections, gelatin desserts, ice cream, and puddings.

Use: Flavoring agent.

Health Effects: A skin irritant in concentrated amounts. GRAS (generally regarded as safe).

Synonyms: CAS: 8000-46-2 ✲ GERANIUM OIL ✲ OIL of GERANIUM ✲ OIL of PELARGONIUM ✲ OIL of ROSE GERANIUM ✲ OIL ROSE GERANIUM ALGERIAN ✲ PELARGONIUM OIL ✲ ROSE GERANIUM OIL ALGERIAN

GERANYL ACETATE

Products: In lavender-scented perfumes, colognes, toiletries, and soaps.

Use: In fragrances, flavoring, and odorants.

Health Effects: Harmless when used for intended purposes.

Synonyms: CAS: 105-87-3 ✲ GERANIOL ACETATE

GERANYL BUTYRATE

Products: In berry and fruit flavorings for beverages, desserts, candy, and bakery products. In rose-scented perfumes and soaps.

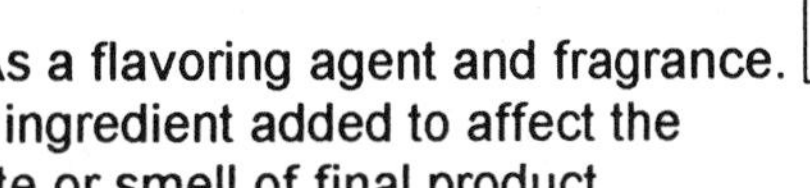

Use: As a flavoring agent and fragrance. An ingredient added to affect the taste or smell of final product.

Health Effects: Harmless when used for intended purposes.

Synonyms: 3,7-DIMETHYL-2,6-OCTADIENE-1-YL BUTYRATE

GERANYL FORMATE

Products: In scented perfumes, toiletries, colognes, and soaps.

Use: As a fruit and floral aroma or fragrance.

Health Effects: Harmless when used for intended purposes.

Synonyms: CAS: 105-86-2 ✲ GERANIOL FORMATE

GERANYL ISOVALERATE

Products: In synthetic berry and fruit flavoring for baked goods, beverages, candy, and ice cream.

Use: Flavoring agent.

Health Effects: A skin irritant in concentrated amounts.

Synonyms: CAS: 109-20-6 ✲ trans-3,7-DIMETHYL-2,6-OCTADIENYL ISOPENTANOATE ✲ (E)-ISOVALERIC ACID-3,7-DIMETHYL-2,6-OCTADIENYL ESTER ✲ (E)-3-METHYLBUTYRIC ACID-3,7-DIMETHYL-2,6-OCTADIENYL ESTER

GERANYL PHENYLACETATE

Products: In honey-rose scented and seasoned items.

Use: A flavoring and odorant agent in foods and fragrances.

Health Effects: Could cause allergic reaction.

Synonyms: 3,7-DIMETHYL-2,6-OCTADIEN-1-YL PHENYLACETATE

GERANYL PROPIONATE

Products: In rose-scented perfumes, colognes, soaps, or fruit flavored foods.

Use: As a fragrance, flavoring, or odorant.

Health Effects: In susceptible individuals it could cause allergic reaction.

Synonyms: 3,7-DIMETHYL-2,6-OCTADADIEN-1-YL PROPIONATE

GHATTI GUM

Products: In butter and butterscotch flavorings. An emulsifier (stabilizes mixes to aid in suspension of liquids) for oils, fats, and waxes.

Use: As a thickener and flavoring.

Health Effects: Could cause allergic reaction.

Synonyms: None found.

GINGER OIL

Products: In baked goods, beverages, desserts, meat, seasonings, relishes, and sauces. In perfumes.

Use: As a food flavoring agent and odorant for colognes.

Health Effects: A skin irritant. Pure substance is a mutagen (changes inherited characteristics) FDA states GRAS (generally regarded as safe) when used for intended purposes.

Synonyms: CAS: 8007-08-7

GLUCONO Δ-LACTONE

Products: In dessert mixes, frankfurters, genoa salamis, meat food products, and sausages.

Use: As an acidifier (to make tart), binder (to hold ingredients together), curing agent, leavening agent, pickling agent, sequestrant (affects the final product appearance or taste).

Health Effects: GRAS (generally regarded as safe) by FDA when used within limits for intended purposes.

Synonyms: CAS: 90-80-2

d-GLUCOSE

Products: In bakery products, baby foods, brewing, wine making, confections, ham (chopped or processed), hamburger, ice cream, luncheon meat, meat loaf, poultry, and sausage.

Use: As a formulation aid, humectant (moisturizer), sweetener (nutritive), texturizing agent (improves feel or texture).

Health Effects: In excessive amounts the concentrated chemical is mildly toxic by swallowing.

Synonyms: CAS: 50-99-7 ❖ CARTOSE ❖ CERELOSE ❖ CORN SUGAR ❖ DEXTROPUR ❖ DEXTROSE ❖ DEXTROSE, anhydrous ❖ DEXTROSOL ❖ GLUCOLIN ❖ GLUCOSE ❖ *d*-GLUCOSE, anhydrous ❖ GLUCOSE LIQUID ❖ GRAPE SUGAR ❖ SIRUP

l-GLUTAMIC ACID

Products: In foods as a flavor enhancer. In tobacco as a taste enhancer. In cosmetics and toiletries as a preservative.

Use: As a dietary supplement, nutrient, antioxidant (slows down spoiling), and a salt substitute.

Health Effects: The effect on the human body by swallowing pure substances are headache, nausea, or vomiting. GRAS (generally regarded as safe) by the FDA when used at moderate levels to accomplish the desired results.

Synonyms: CAS: 56-86-0 ❖ α-AMINOGLUTARIC ACID ❖ *l*-2-AMINOGLUTARIC ACID ❖ 2-AMINOPENTANEDIOIC ACID ❖ 1-AMINOPROPANE-1,3-DICARBOXYLIC ACID ❖ GLUSATE ❖ GLUTACID ❖ GLUTAMIC ACID ❖ α-GLUTAMIC ACID ❖ *d*-GLUTAMIENSUUR ❖ GLUTAMINIC ACID ❖ *l*-GLUTAMINIC ACID ❖ GLUTAMINOL ❖ GLUTATON

l-GLUTAMIC ACID HYDROCHLORIDE

Products: In beer as a flavor enhancer. In cosmetics as a preservative. In hair permanent products.

Use: As a dietary supplement, flavoring agent, nutrient, salt substitute, or antioxidant (slows down spoiling).

Health Effects: A general purpose additive that FDA states is harmless when used for the intended purposes within limitations.

Synonyms: CAS: 138-15-8 ❖ α-AMINOGLUTARIC ACID HYDROCHLORIDE ❖ *l*-2-AMINOGLUTARIC ACID HYDROCHLORIDE ❖ 2-AMINOPENTANEDIOIC ACID HYDROCHLORIDE ❖ 1-AMINOPROPANE-1,3-DICARBOXYLIC ACID HYDROCHLORIDE ❖ GLUTAMIC ACID HYDROCHLORIDE ❖ α-GLUTAMIC ACID HYDROCHLORIDE ❖ GLUTAMINIC ACID HYDROCHLORIDE ❖ *l*-GLUTAMINIC ACID HYDROCHLORIDE

GLUTAMINE

Products: Various.

Use: As a dietary supplement and nutrient.

Health Effects: In gross amounts it is mildly toxic by swallowing. FDA approves use within limitations.

Synonyms: CAS: 56-85-9 ❖ 2-AMINOGLUTARAMIC ACID ❖ *l*-2-AMINOGLUTARAMIDIC ACID ❖ CEBROGEN ❖ GLUMIN ❖ GLUTAMIC ACID AMIDE ❖ GLUTAMIC ACID-5-AMIDE ❖ γ-GLUTAMINE ❖ *l*-GLUTAMINE ❖ LEVOGLUTAMID ❖ LEVOGLUTAMIDE ❖ STIMULINA

GLYCERIN

Products: In baked goods, candy, marshmallows, tobacco, jelly beans, ice cream toppings, sauces, cosmetics, blushes, skin creams, hair spray, mouth wash, and toothpaste.

Use: A food additive, humectant (keeps products moist), plasticizer (in coatings of cheeses and sausages), and solvent.

Health Effects: Mildly toxic by swallowing large amounts of pure substance. Effects on the human body by swallowing are: headache, nausea, or vomiting. A human mutagen (changes inherited characteristics). A skin and eye irritant. In the form of mist it is a breathing irritant. GRAS (generally regarded as safe) by FDA when used in limited amounts for intended purposes.

Synonyms: CAS: 56-81-5 ✲ GLYCERIN, anhydrous ✲ GLYCERINE ✲ GLYCERIN, synthetic ✲ GLYCERITOL ✲ GLYCEROL ✲ GLYCYL ALCOHOL ✲ GROCOLENE ✲ MOON ✲ 1,2,3-PROPANETRIOL ✲ STAR ✲ SUPEROL ✲ SYNTHETIC GLYCERIN ✲ 90 TECHNICAL GLYCERINE ✲ TRIHYDROXYPROPANE ✲ 1,2,3-TRIHYDROXYPROPANE

GLYCEROL ESTER of PARTIALLY DIMERIZED ROSIN

Products: In chewing gum, confectionery, eggs (shell), tablet form food supplements, fruits, gum, and vegetables.

Use: As a color diluent; masticatory substance in chewing gum base.

Health Effects: FDA approves use in moderate amounts to accomplish the intended effects.

Synonyms: None found.

GLYCEROL ESTER of POLYMERIZED ROSIN

Products: In chewing gum and gumballs.

Use: As masticatory substance in gum base.

Health Effects: FDA approves use in moderate amounts to accomplish the desired results.

Synonyms: None found.

GLYCEROL ESTER of WOOD ROSIN

Products: In beverages (citrus) and chewing gum.

Use: As a beverage stabilizer (maintains a uniform consistency), masticatory substance in gum base.

Health Effects: Harmless when used in limited amounts for intended purposes.

Synonyms: None found.

GLYCEROL-LACTO PALMITATE

Products: In cakes, fats (rendered animal), and whipped topping.

Use: As an emulsifier (it stabilizes mixture and keeps it uniform).

Health Effects: FDA approves use in moderate amounts for intended purposes.

Synonyms: None found.

GLYCEROL-LACTO STEARATE

Products: In cake mixes, chocolate coatings, fats (rendered animal), shortening, and whipped vegetable toppings.

Use: As an emulsifier (it stabilizes mixture and keeps it uniform).

Health Effects: Harmless when used for intended purposes.

Synonyms: None found.

GLYCEROL MONOOLEATE

Products: In coffee whiteners, packaging materials, and vegetable oil.

Use: As a defoamer, dispersing agent, emulsifier (it stabilizes mixture and keeps it uniform), and plasticizer.

Health Effects: FDA approves use at moderate levels to accomplish the desired results.

Synonyms: None found.

GLYCERYL MONOOLEATE

Products: In baking mixes, beverages (nonalcoholic), beverage bases, chewing gum, and meat products.

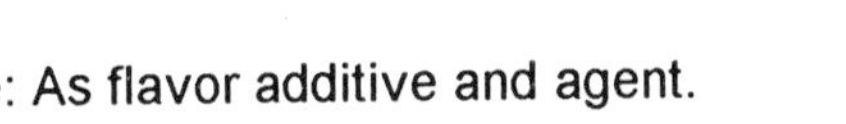

Use: As flavor additive and agent.

Health Effects: Harmless when used for intended purposes.

Synonyms: CAS: 25496-72-4

GLYCERYL TRISTEARATE

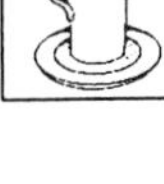

Products: In chocolate (imitation), cocoa, confections, fats, and oil.

Use: As a crystallization accelerator, fermentation aid, formulation aid, fractionation aid, lubricant, release agent, surface-finishing agent, and winterization agent.

Health Effects: FDA approves use within stated limits to accomplish the desired results.

Synonyms: CAS: 555-43-1

GLYCINE

Products: In beverage bases, beverages, fats (rendered animal), and cosmetic texturizers.

Use: A nutrient; it reduces the bitter taste of saccharin and prevents fats from getting rancid,

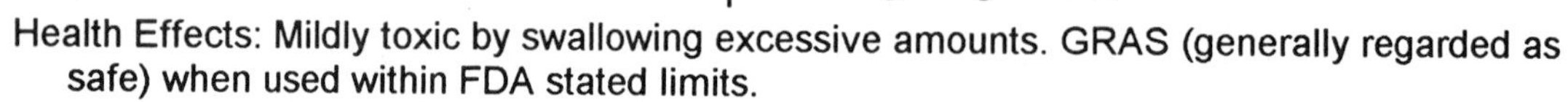

Health Effects: Mildly toxic by swallowing excessive amounts. GRAS (generally regarded as safe) when used within FDA stated limits.

Synonyms: CAS: 56-40-6 ❋ AMINOACETIC ACID ❋ GLYCOLIXIR ❋ HAMPSHIRE GLYCINE

GLYCYRRHIZIN

Products: Extracted from licorice root; extremely sweet. Tobacco humectant (prevents drying out), confectionery, root beer foaming agent, chocolate, cocoa, and chewing gum. Masks taste in pharmaceuticals such as aspirin.

Use: As a sweetener, nutritive, and flavoring.

Health Effects: Harmless when used for intended purposes.

Synonyms: None found.

GOLD

Products: In dental alloys, jewelry, industrial products, and medical applications.

Use: Various.

Health Effects: Harmless when used for intended purposes. Frequently suspected of causing allergic reaction when the actual cause is the alloy metal such as nickel.

Synonyms: CAS: 7440-57-5

GRAPEFRUIT OIL

Products: In bakery products, beverages (nonalcoholic), chewing gum, confections, gelatin desserts, ice cream products, puddings, and syrups.

Use: As a flavoring agent.

Health Effects: A skin irritant in concentrated amounts. GRAS (generally regarded as safe) when used in moderate levels to accomplish the desired results.

Synonyms: CAS: 8016-20-4 ❋ GRAPEFRUIT OIL, coldpressed ❋ GRAPEFRUIT OIL, expressed ❋ OIL of GRAPEFRUIT ❋ OIL of SHADDOCK

GRAPE SKIN EXTRACT

Products: In beverages (alcoholic), beverages (carbonated), and beverages (still).

Use: As a coloring agent and fruit taste enhancer.

Health Effects: Harmless when used for intended purposes. Could cause allergic reactions in some susceptible individuals.

Synonyms: ENOCIANINA

GRAPHITE

Products: In bricks, shingles, lubricants, paints, coatings, pencil lead, makeup pencils, and cosmetic products.

Use: As polish, lubricant, filler, and cement in industrial and consumer products.

Health Effects: In powder form it is a fire risk.

Synonyms: CAS: 7782-42-5 ✲ BLACK LEAD ✲ PLUMBAGO

GUANIDINE CARBONATE

Products: In soap and cosmetic products.

Use: As a moisturizer.

Health Effects: Toxic by swallowing.

Synonyms: None found.

GUANINE

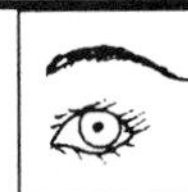

Products: In pearl finish nail polish, eye shadow, and lipstick. Replaced by synthetic chemicals such as bismuth.

Use: Pearly coloration essence for cosmetics and toiletries.

Health Effects: Harmless when used for intended purposes. It is derived from guano, sugar beets, yeast, clover seed, and fish scales.

Synonyms: CAS: 73-40-5 ✲ 2-AMINO-6-OXYPURINE ✲ 2-AMJNOHYPOXANTHINE ✲ MEARLMAID

GUANYLIC ACID SODIUM SALT

Products: In canned foods, poultry, sauces, snack items, and soups.

Use: As a flavor enhancer.

Health Effects: In concentrated amounts it is mildly toxic by swallowing large amounts.

Synonyms: CAS: 5550-12-9 ✲ DISODIUM GMP ✲ DISODIUM-5′-GMP ✲ DISODIUM GUANYLATE (FCC) ✲ DISODIUM-5′-GUANYLATE ✲ GMP DISODIUM SALT ✲ 5′-GMP DISODIUM SALT ✲ GMP SODIUM SALT ✲

SODIUM GMP ✲ SODIUM GUANOSINE-5′-MONOPHOSPHATE ✲ SODIUM GUANYLATE ✲ SODIUM-5′-GUANYLATE

GUAR GUM

Products: In baked goods, baking mixes, beverages, cereals (breakfast), cheese, dairy product analogs, fats, gravies, ice cream, jams, jellies, milk products, oil, sauces, soup mixes, soups, sweet sauces, syrups, toppings, vegetable juices, vegetables (processed), and cheese spreads. Also cosmetic products and creams.

Use: As an emulsifier, firming agent, formulation aid, stabilizer, and thickening agent.

Health Effects: It is mildly toxic by swallowing large amounts of pure material. GRAS (generally regarded as safe) when used within FDA limits.

Synonyms: CAS: 9000-30-0 ✲ A-20D ✲ BURTONITE V-7-E ✲ CYAMOPSIS GUM ✲ DEALCA TP1 ✲ DECORPA ✲ GALACTASOL ✲ GENDRIV 162 ✲ GUAR ✲ GUAR FLOUR ✲ GUM CYAMOPSIS ✲ GUM GUAR ✲ INDALCA AG ✲ JAGUAR No. 124 ✲ JAGUAR GUM A-20-D ✲ JAGUAR PLUS ✲ LYCOID DR ✲ REGONOL ✲ REIN GUARIN ✲ SUPERCOL U POWDER

GUM GHATTI

Products: In beverage mixes, beverages, buttered syrup, and oil.

Use: An emulsifier (stabilizes and maintains mixes).

Health Effects: Mildly toxic by swallowing large amounts of pure substance. GRAS (generally regarded as safe) when used within FDA limits.

Synonyms: CAS: 9000-28-6 ✲ INDIAN GUM

GUM GUAIAC

Products: Fats (rendered animal).

Use: As an antioxidant (slows down spoiling) and preservative.

Health Effects: Moderately toxic by swallowing gross amounts of concentrated material. Harmless when used within limits for intended purposes.

Synonyms: GUAIAC GUM

"A disposable diaper or polystyrene trash bag may legally be called "biodegradable" even if it takes 1,000 years in a landfill to disintegrate."

Good Housekeeping Magazine

HALLUCINOGEN

Products: Natural plants, synthetic narcotics, and alkaloids.

Use: For illegal, addictive, hallucinatory effects.

Health Effects: Addictive drugs which act on the body and cause mental disturbance, imaginary disturbances, imaginary experiences, coma, and possibly death. Sale and possession (except by physicians) is illegal in U. S.

Synonyms: CANNABIS ✲ MARIJUANA ✲ HASHISH ✲ LYSERGIC ACID (LAD) ✲ AMPHETAMINES ✲ MORPHINE DERIVATIVES

HELIUM

Products: For welding, balloon inflation (weather, research, party), diver's breathing equipment, and inflated advertising signs.

Use: For inflating, breathing, and pressurizing.

Health Effects: A simple asphyxiant. Nonflammable gas.

Synonyms: CAS: 7440-59-7 ✲ HELIUM, compressed (DOT) ✲ HELIUM, refrigerated liquid (DOT)

HENNA

Products: In hair dyeing products. In antifungal medications.

Use: As a red coloring agent.

Health Effects: Always follow directions carefully when using products containing this item. Derived from the dried leaves of tropical plants.

Synonyms: Not found.

HEMATOXYLIN

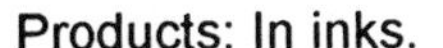

Products: In inks.

Use: As a colorant.

Health Effects: A possible carcinogen (may cause cancer).

Synonyms: CAS: 517-28-2

HEPTACHLOR

Products: Pesticide for insects.

Use: Termiticide (kills termites).

Health Effects: Suspected carcinogen (may cause cancer). A poison by swallowing and skin contact. A human mutagen (changes inherited characteristics). Acute exposure and chronic doses have caused liver damage. See also closely related chlordane. In humans, a dose of 1-3 grams can cause serious symptoms, especially where liver impairment is involved. Symptoms include tremors, convulsions, kidney damage, respiratory collapse, and death. EPA has canceled use of pesticides containing heptachlor except for exterior below grade application, for termiticide.

Synonyms: CAS: 76-44-8 ✲ AGROCERES ✲ 3-CHLOROCHLORDENE ✲ DRINOX ✲ 3,4,5,6,7,8,8-HEPTACHLORODICYCLOPENTADIENE ✲ HEPTAGRAN ✲ HEPTAMUL ✲ RHODIACHLOR

n-HEPTADECANOL

Products: In perfumes, shaving products, shaving lotions, colognes, soaps, toiletries, and cosmetics. In the manufacture of detergents.

Use: As plasticizer, fixative, and wetting agent.

Health Effects: Harmless when used for intended purposes.

Synonyms: Not found.

HEPTANAL

Products: In perfumes, fragrances, foods, beverages, ice creams, candy, and bakery products.

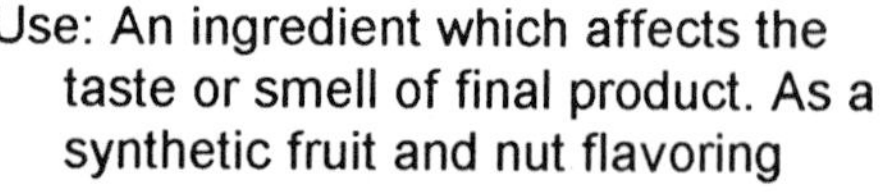

Use: An ingredient which affects the taste or smell of final product. As a synthetic fruit and nut flavoring

Health Effects: Mildly toxic by swallowing large, concentrated amounts.

Synonyms: CAS: 111-71-7 ✲ ENANTHAL ✲ ENANTHALDEHYDE ✲ ENANTHOLE ✲ HEPTALDEHYDE ✲ OENANTHALDEHYDE ✲ OENANTHOL

n-HEPTANOIC ACID

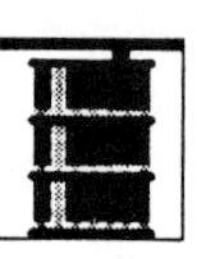

Products: In lubricants and brake fluids.

Use: For slipperiness and reducing friction.

Health Effects: Mildly toxic by swallowing.

Synonyms: CAS: 111-14-8 ✲ ENANTHIC ACID ✲ n-HEPTYLIC ACID ✲ HEPTOIC ACID

HEPTYL ACETATE

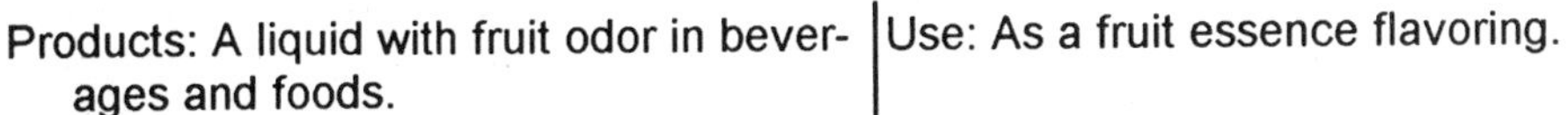

Products: A liquid with fruit odor in beverages and foods.

Use: As a fruit essence flavoring.

Health Effects: Irritating to skin, eyes, nose, and throat in pure form. Harmless when used for intended purposes.

Synonyms: CAS: 112-06-1 ✲ HEPTANYL ACETATE ✲ ACETATE C-7 ✲ 1-HEPTYL ACETATE

HEPTYL ALCOHOL

Products: In cosmetics and perfumery.

Use: For formulating and as a solvent.

Health Effects: Moderately toxic by swallowing and excessive skin contact.

Synonyms: CAS: 111-70-6 ✲ 1-HEPTANOL ✲ ENANTHYL ALCOHOL

HEPTYL FORMATE

Products: In baked goods, beverages, candy, and ice cream.

Use: As artificial fruit-essence flavoring.

Health Effects: FDA approves use at moderate levels to accomplish the intended effect. A skin irritant in concentrated amounts.

Synonyms: CAS: 112-23-2 ✲ FORMIC ACID, HEPTYL ESTER ✲ HEPTANOL, FORMATE ✲ n-HEPTYL METHANOATE

HEPTYL HEPTOATE

Products: In food and beverage products.

Use: As artificial fruit-essence flavoring.

Health Effects: Harmless when used for intended purposes.

Synonyms: None found.

HEPTYL PARABEN

Products: In beer, beverages (fermented malt), fruit drinks (noncarbonated), wine, and soft drinks (noncarbonated).

Use: An antioxidant (maintains freshness) or preservative.

Health Effects: Harmless when used for intended purposes within designated limits.

Synonyms: n-HEPTYL p-HYDROXYBENZOATE ✲ HEPTYL ESTER OF PARA-HYDROXYBENZOIC ACID

HEPTYL PELARGONATE

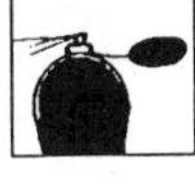

Products: In foods, beverages, colognes,

Use: For flavoring and fragrancing. An

and perfumes.

ingredient which is added to affect the taste or smell of final product.

Health Effects: Harmless when used for intended purposes.

Synonyms: None found.

HESPERIDIN

Products: In assorted food and beverage items.

Use: As a synthetic sweetener.

Health Effects: Extracted from citrus fruit peel. Can cause allergic reaction.

Synonyms: CAS: 520-26-3 ❊ CIRANTIN ❊ VITAMIN P

HEXACHLOROBENZENE

Products: On outdoor furniture, fences, foundations, and landscape timbers.

Use: A wood preservative.

Health Effects: A carcinogen (causes cancer). Toxic by swallowing.

Synonyms: CAS: 118-74-1 ❊ PERCHLOROBENZENE

HEXACHLOROPHENE

Products: In germicidal soaps and foot powders.

Use: A bactericide once utilized in many cleansing products.

Health Effects: FDA prohibits use unless prescribed by a physician.

Synonyms: CAS: 70-30-4

HEXAHYDROBENZOIC ACID

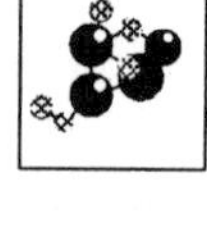

Products: In paints, varnishes, dry cleaning soaps, oils, and detergents.

Use: As a drier, solvent, lubricant, and stabilizer

Health Effects: Label directions should be followed very carefully for products which contain this chemical.

Synonyms: CYCLOHEXANECARBOXYLIC ACID ❊ NAPTHENIC ACID

HEXAMETHYLENETRAMINE

Products: In adhesives (rubber to textile), fungicide, textiles (shrink-proofing), and antibacterial.

Use: Various.

Health Effects: A skin irritant.

Synonyms: CAS: 100-97-0 ✲ METHENAMINE ✲ HMTA ✲ AMINOFORM ✲ HEXAMINE

1-HEXANAL

Products: A fruit and coconut flavoring in beverages, desserts, candies, bakery products, and gels. In bactericides and perfumes.

Use: As flavoring and fragrancing.

Health Effects: Concentrated form is mildly toxic by swallowing large amounts and by breathing fumes. An irritant to skin and eyes. FDA approves use at moderate levels to accomplish the intended effect.

Synonyms: CAS: 66-25-1 ✲ ALDEHYDE C-6 ✲ CAPROALDEHYDE ✲ CAPROIC ALDEHYDE ✲ CAPRONALDEHYDE ✲ n-CAPROYLALDEHYDE ✲ HEXALDEHYDE ✲ HEXANAL

1,2,6-HEXANETRIOL

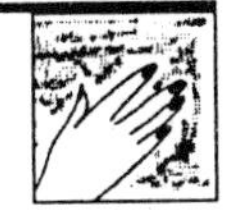

Products: In skin lotions and cosmetic makeup.

Use: As a softener and moisturizer.

Health Effects: Mildly toxic by swallowing. An eye and skin irritant.

Synonyms: CAS: 106-69-4 ✲ HEXANE-1,2,6 TRIOL ✲ HEXANETRIOL-1,2,6

HEXAPHOS

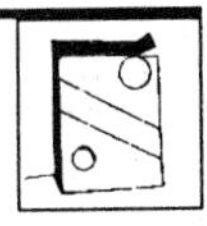

Products: In cleansers, laundry soaps, dishwashing powders, and detergents.

Use: As a water-softener and soap deposit preventative.

Health Effects: Follow label directions carefully on products containing this chemical.

Synonyms: GLASSY PHOSPHATE

HEXAZINONE

Products: In herbicides.

Use: As a weed killer.

Health Effects: Moderately toxic by swallowing. Mildly toxic by skin contact. An eye irritant. USDA rates it as a minimal hazard.

Synonyms: CAS: 51235-04-2 ✲ 3-CYCLOHEXYL-6-(DIMETHYLAMINO)-1-METHYL-s-TRIAZINE-2,4(1H,3H)-DIONE ✲ 3-CYCLOHEXYL-6-(DIMETHYLAMINO)-1-METHYL-1,3,5-TRIAZINE-2,4(1H,3H)-DIONE ✲ DPX 3674 ✲ VELPAR ✲ VELPAR WEED KILLER

HEXENOL

Products: Occurs naturally in grasses, leaves, herbs, and teas. In perfumes

Use: A fragrancing or scenting additive.

and odorants.

Health Effects: Could cause allergic reaction. Harmless when used for intended purposes.

Synonyms: 3-HEXEN-1-OL ✲ LEAF ALCOHOL

HEXETIDINE

Products: In bactericides, fungicides, algicides; also for synthetic textiles.

Use: As antifungal (kills fungus), antibacterial (kills bacteria), algicidal (kills algea), and antistatic agent.

Health Effects: Harmless when directions are followed and product is used for intended purposes.

Synonyms: CAS: 141-94-6 ✲ (AMINO-1,3-BIS[β-ETHYLHEXYL]-5-METHYLHEXAHYDROPYRIMIDINE

HEXYL ACETATE

Products: A synthetic fruit and berry taste enhancer for beverages, desserts, candies, bakery products, and flavored gum.

Use: As a flavoring agent.

Health Effects: In large amounts it is mildly toxic by swallowing pure chemical. FDA approves use at moderate levels to accomplish the desired results.

Synonyms: CAS: 142-92-7 ✲ ACETIC ACID HEXYL ESTER ✲ n-HEXYL ACETATE ✲ 1-HEXYL ACETATE ✲ HEXYL ALCOHOL, ACETATE ✲ HEXYL ETHANOATE

HEXYL ALCOHOL

Products: In pharmaceuticals, antiseptics, perfumes, toiletries, and textiles.

Use: A plasticizer, solvent, intermediate, and finishing agent.

Health Effects: Moderately toxic by swallowing, and by excessive skin contact. A skin and severe eye irritant.

Synonyms: CAS: 111-27-3 ✲ 1-HEXANOL ✲ AMYL CARBINOL

HEXYLENE

Products: In flavors, perfumes, and dyes.

Use: Various.

Health Effects: Moderately toxic irritant to skin, eyes, nose, and throats. Could cause allergic reaction.

Synonyms: CAS: 592-41-6 ✲ 1-HEXENE

HEXYLENE GLYCOL

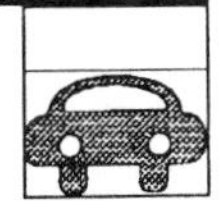

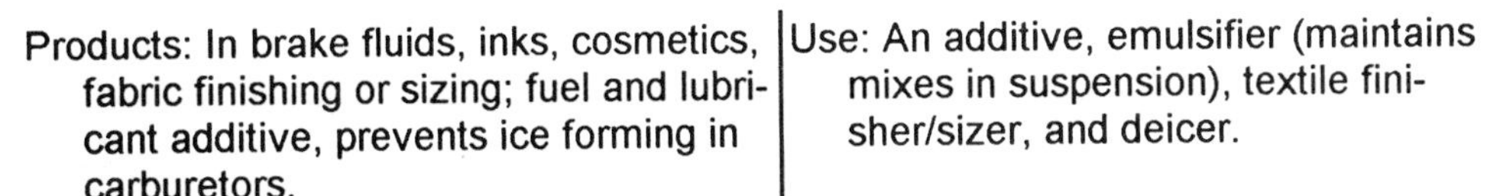

Products: In brake fluids, inks, cosmetics, fabric finishing or sizing; fuel and lubricant additive, prevents ice forming in carburetors.

Use: An additive, emulsifier (maintains mixes in suspension), textile finisher/sizer, and deicer.

Health Effects: Moderately toxic by swallowing. Mildly toxic by skin contact. Effects on the human body by breathing are conjunctiva (eye), olfactory (nasal), and pulmonary (lung) changes.

Synonyms: CAS: 107-41-5 ✲ 4-METHYL-2,4-PENTANEDIOL

HEXYLRESORCINOL

Products: In medications and pharmaceuticals.

Use: An antibacterial and anthelmintic (destroys and expels worms).

Health Effects: Irritant to respiratory tract, eye, and skin. Concentrated solutions can cause burns of the skin, nose, and throats. A topical antiseptic. In worming medicine (it destroys worms).

Synonyms: CAS: 136-77-6 ✲ 1,3-DIHYDROXY-4-HEXYLBENZENE

HEXYL CINNAMALDEHYDE

Products: A jasmine-type odor for fruit, berry, beverage honey flavorings, desserts, candies, and bakery products.

Use: As a flavoring, aroma, and taste enhancer.

Health Effects: Moderately toxic by swallowing gross, concentrated amounts. A skin irritant in concentration. Harmless when used for intended effects.

Synonyms: CAS: 101-86-0 ✲ α-HEXYLCINNAMALDEHYDE ✲ HEXYL CINNAMIC ALDEHYDE ✲ α-HEXYLCINNAMIC ALDEHYDE ✲ α-n-HEXYL-β-PHENYLACROLEIN ✲ 2-(PHENYLMETHYLENE)OCTANOL

HIGH-FRUCTOSE CORN SYRUP

Products: In beverages (carbonated), candy, confections, desserts (frozen), drinks (dairy), fruits (canned), ham (chopped or processed), hamburger, ice cream, luncheon meat, meat loaf, poultry, and sausage.

Use: As a flavoring or sweetening (nutritive) agent.

Health Effects: GRAS (generally regarded as safe). Harmless when used for intended purposes.

Synonyms: CORN SYRUP, HIGH-FRUCTOSE

HISTIDINE

Products: Medicine. In cosmetic skin products. A dietary supplement and nutrient.

Use: Various.

Health Effects: FDA approves use at moderate levels to accomplish the intended effect when used within stated limitations. A mutagen (changes inherited characteristics).

Synonyms: CAS: 71-00-1 ✻ *l*-α-AMINO-4(OR 5)-IMIDAZOLEPROPIONIC ACID ✻ GLYOXALINE-5-ALANINE ✻ *l*-HISTIDINE

HOMOMENTHYL SALICYLATE

Products: In suntan lotions, creams, oils, toiletries, protectants, and cosmetics.

Use: To filter or absorb UV (ultraviolet) radiation of sunlight.

Health Effects: Could cause allergic reaction. Harmless when used for intended purposes.

Synonyms: 3,3,5-trimethylcyclohexyl salicylate

HOMOSALATE

Products: In suntan lotions, creams, oils, protectants, and cosmetics.

Use: A UV (ultraviolet) filter/screen.

Health Effects: Could cause allergic reaction. Harmless when used for intended purposes.

Synonyms: None found.

HOP EXTRACT, MODIFIED

Products: For beer, fruit, and root beer.

Use: As a flavoring agent for beverages.

Health Effects: Harmless when used for intended purposes.

Synonyms: MODIFIED HOP EXTRACT.

HOPS OIL

Products: In berry, whiskey, beverage, desserts, bakery products, and seasonings.

Use: As an aromatic, spicy, bitter, beverage flavoring.

Health Effects: GRAS (generally regarded as safe) when used at a moderate level to accomplish the intended effects.

Synonyms: None found.

HUMIC ACID

Products: For printing ink pigments. Cosmetic facial mud baths.

Use: As coloring, softening, moisturizing, agent.

Health Effects: Harmless when used for intended purposes. It is a natural stream pollutant and is thought to be capable of triggering the "red tide" phenomenon due to microorganisms in seawater.

Synonyms: None found.

HYDRATED ALUMINUM OXIDES

Products: In paper, polishes, adhesives, inks, paints, and cosmetics.

Use: As a coloring agent.

Health Effects: Harmless when used for intended purposes.

Synonyms: None found.

HYDRAPHTHAL

Products: For fabrics and textiles.

Use: A solvent and detergent.

Health Effects: Harmless when used for intended purposes.

Synonyms: None found.

HYDRARGAPHEN

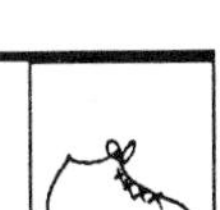

Products: For wool, adhesives, fabrics, leather, paints, and wood products.

Use: A germicide, bactericide, and fungicide.

Health Effects: A poison. A severe eye irritant.

Synonyms: CAS: 14235-86-0 ✲ HYDRAPHEN ✲ CONOTRANE ✲ PENOTRANE ✲ SEPTOTAN

HYDRAZINE SULFATE

Products: Fungicides and germicides.

Use: Destroys fungi and bacteria.

Health Effects: A carcinogen (causes cancer). A poison by swallowing. Effects on the human body by swallowing are paresthesia (abnormal sensations), Somnolence (sleepiness), nausea, or vomiting. A mutagen (changes inherited characteristics). An eye irritant.

Synonyms: CAS: 10034-93-2 ✲ DIAMINE SULFATE ✲ DIAMIDOGEN SULFATE

HYDRIODIC ACID

Products: An expectorant (aids in loosening secretions) in cough syrups. A disinfectant.

Use: As medication and as a germicide.

Health Effects: In large amounts it is poison by swallowing and breathing concentrated fumes. A corrosive and poisonous irritant to skin, eyes, nose, and throats.

Synonyms: CAS: 10034-85-2 ✲ HYDROGEN IODIDE

HYDROCHLORIC ACID

Products: In hair bleaching products, swimming pool chemicals, toilet cleaner, drain cleaner, a buffer, a neutralizing agent, and an etching agent.

Use: An oxidizer in hair rinses and color remover in hair products. Also in food processing.

Health Effects: A human poison. Mildly toxic to humans by breathing fumes. A corrosive irritant to the skin, eyes, nose, and throats. A concentration of 35 ppm causes irritation of the throat after short exposure. On EPA Extrememly Hazardous Substance list.

Synonyms: CAS: 7647-01-0 ✲ CHLOROHYDRIC ACID ✲ HYDROCHLORIC ACID, anhydrous ✲ HYDROCHLORIC ACID, solution, inhibited ✲ HCL ✲ HYDROGEN CHLORIDE ✲ HYDROGEN CHLORIDE, anhydrous ✲ HYDROGEN CHLORIDE, refrigerated liquid ✲ MURIATIC ACID ✲ SPIRITS of SALT

HYDROCINNAMIC ACID

Products: In perfumes and fragrances. An ingredient to affect the taste or smell of final products.

Use: As flavoring and scenting agent.

Health Effects: Moderately toxic by swallowing excessive amounts. Mildly toxic by skin contact. A skin irritant. FDA approves use at a moderate levels to accomplish the intended effect.

Synonyms: CAS: 122-97-4 ✲ 3-BENZENEPROPANOL ✲ HYDROCINNAMYL ALCOHOL ✲ (3-HYDROXYPROPYL)BENZENE ✲ γ-PHENYLPROPANOL ✲ 3-PHENYLPROPANOL ✲ PHENYLPROPYL ALCOHOL ✲ γ-PHENYLPROPYL ALCOHOL ✲ 3-PHENYLPROPYL ALCOHOL

HYDROCOLLOID

Products: In food and cosmetic products for emulsifiers, thickeners, and gelling agents.

Use: Imparts smoothness and texturizing to products.

Health Effects: Harmless when used for intended purposes.

Synonyms: GUM ARABIC ✲ AGAR ✲ GUAR GUM ✲ STARCHES ✲ DEXTRAN ✲ GELATIN

HYDROCYANIC ACID

Products: In rodenticides and pesticides.

Use: As rat poison and insect killer.

Health Effects: Toxic by swallowing, breathing, and skin absorption.

Synonyms: CAS: 74-90-8 ✲ PRUSSIC ACID ✲ HYDROGEN CYANIDE ✲ FORMONITRILE

HYDROGENATED VEGETABLE OIL

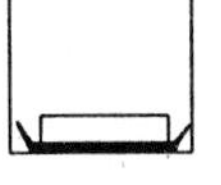

Products: Margarine, shortening, and many processed foods.

Use: As flavoring, seasoning, cooking, and frying oil.

Health Effects: Diets high in fats and oils cause heart disease, obesity, and possibly cancer.

Synonyms: None found.

HYDROGEN PEROXIDE

Products: In cheese whey (annatto-colored), corn syrup, distilling materials, eggs (dried), emulsifiers containing fatty acid esters, herring, milk, packaging materials, starch, tea (instant), tripe, whey, wine, and wine vinegar. Also bleaching products for skin and hair. In cosmetic creams, mouthwashes, dentifrices, and hair wave products.

Use: An antiseptic, bleaching, oxidizing agent, and preservative.

Health Effects: In concentration it is moderately toxic by breathing fumes, swallowing large amounts, and by skin contact. A corrosive irritant to skin, eyes, nose, and throats. A human mutagen (changes inherited characteristics). Questionable carcinogen (causes cancer). FDA approves use within limitations.

Synonyms: CAS: 7722-84-1 ✲ ALBONE ✲ DIHYDROGEN DIOXIDE ✲ HIOXYL ✲ HYDROGEN DIOXIDE ✲ HYDROGEN PEROXIDE, solution (over 52% peroxide) ✲ HYDROGEN PEROXIDE, stabilized (over 60% peroxide) ✲ HYDROPEROXIDE ✲ INHIBINE ✲ OXYDOL ✲ PERHYDROL ✲ PERONE ✲ PEROXAN ✲ PEROXIDE ✲ SUPEROXOL

HYDROLYZED VEGETABLE PROTEIN

Products: In dried soup mixes, beef products, frankfurters, sauce mixes, and canned stews.

Use: A flavoring, seasoning, and taste enhancing agent.

Health Effects: Could cause allergic reactions. GRAS (generally regarded as safe).

Synonyms: HPP ❊ HVP

HYDROQUINONE

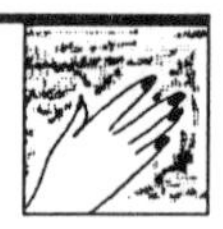

Products: Cosmetic skin bleaching creams, paints, varnishes, fuels, oils, and photographic developers.

Use: As medicated skin lightener and various other uses.

Health Effects: A human poison by swallowing. An active allergen and a strong skin irritant. The swallowing of 1 gram may cause nausea, dizziness, a sense of suffocation, increased respiration, vomiting, pallor, muscle twitching, headache, delirium, and collapse.

Synonyms: CAS: 123-31-9 ❊ QUINOL ❊ HYDROQUINOL ❊ p-DIHYDROXYBENZENE

HYDROQUINONE DIMETHYL ETHER

Products: In paints, perfumes, dyes, cosmetics, and suntan products.

Use: As a weathering agent, fixative, and flavoring (pleasant, sweet, clover odor).

Health Effects: Poison by swallowing. Human reproductive effects (infertility, or sterility, or birth defects).

Synonyms: CAS: 654-42-2 ❊ DMB ❊ 1,4-DIMETHOXYBENZENE ❊ DIMETHYL HYDROQUINONE

HYDROXYACETIC ACID

Products: As fabric dye, leather dye, adhesives, cleaning compound, and polishing compound.

Use: For coloring, finishing, and adhesion.

Health Effects: Harmless when used for intended purposes.

Synonyms: GLYCOLIC ACID

m-HYDROXYBENZALDEHYDE

Products: For beverages, ices, desserts, candies, bakery products, spices and liqueurs. In disinfectants, fumigants, and perfumery.

Use: As a buttery, nutlike, seasoning. Anti-inflammatory properties.

Health Effects: Moderately toxic in pure form by swallowing and by skin contact. A skin irritant. Could cause allergic reaction.

Synonyms: CAS: 90-2-8 ❊ SALICYLALDEHYDE ❊ SALICYLIC ALDEHYDE ❊ 2-FORMYLPHENOL ❊ SALICYLAL

p-HYDROXYBENZOIC ACID

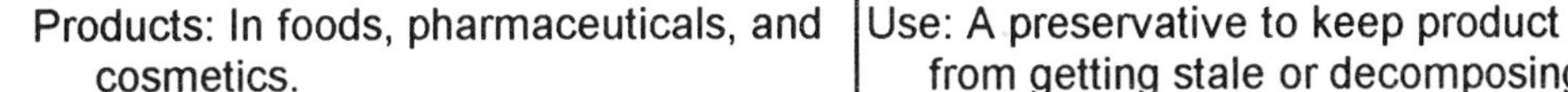

Products: In foods, pharmaceuticals, and cosmetics.

Use: A preservative to keep product from getting stale or decomposing.

Health Effects: Harmless when used for intended purposes. Approved by FDA.

Synonyms: CAS: 99-96-7 ❊ 4-CARBOXYPHENOL ❊ p-HYDROXYBENZOIC ACID ❊ 4-HYDROXYBENZOIC ACID

p-HYDROXYBENZOIC ACID ETHYL ESTER

Products: In baked goods, beverages, food colors, and wine.

Use: As an antimicrobial agent and preservative.

Health Effects: In large, concentrated amounts it is moderately toxic by swallowing. GRAS (generally regarded as safe) when used within limitations stated by FDA.

Synonyms: CAS: 120-47-8 ❊ ASEPTOFORM E ❊ BONOMOLD OE ❊ p-CARBETHOXYPHENOL ❊ EASEPTOL ❊ ETHYL-p-HYDROXYBENZOATE ❊ ETHYL PARABEN ❊ ETHYL PARASEPT ❊ p-HYDROXYBENZOIC ETHYL ESTER

p-HYDROXYBENZOIC ACID METHYL ESTER

Products: In baked goods, beverages, food colors, milk, wine, and cosmetics.

Use: An antimicrobial agent and a preservative.

Health Effects: Moderately toxic by swallowing great amounts of pure chemical. A mutagen (changes inherited characteristics). GRAS (generally regarded as safe) when used within limitations stated by FDA.

Synonyms: CAS: 99-76-3 ❊ ABIOL ❊ ASEPTOFORM ❊ MASEPTOL ❊ METHYLBEN ❊ METHYL CHEMOSEPT ❊ METHYL ESTER of p-HYDROXYBENZOIC ACID ❊ METHYL p-HYDROXYBENZOATE ❊ METHYL p-OXYBENZOATE ❊ METHYLPARABEN ❊ METHYL PARAHYDROXYBENZOATE ❊ METHYL PARASEPT ❊ METOXYDE ❊ MOLDEX ❊ PARABEN ❊ PARASEPT ❊ PARIDOL ❊ SEPTOS

HYDROXYCITRONELLAL

Products: In perfumes, soaps, toiletries, and cosmetics.

Use: A sweet, lily-type, fragrant odor.

Health Effects: Derived from plant *Java citronella* or *Eucalyptus citriodora*. Harmless when used for intended purposes.

Synonyms: CAS: 107-75-5 ❊ CITRONELLAL HYDRATE ❊ 3,7-DIMETHYL-7-HYDROXYACTENAL

HYDROXYCITRONELLAL DIMETHYL ACETAL

Products: In foods and perfumery.

Use: As a flavoring agent and odorant.

Health Effects: Harmless when used for intended purposes.

Synonyms: HYDROXYACETAL

2-HYDROXY ETHYL CARBAMATE

Products: For wash and wear cotton fabrics.

Use: As a sizing and finishing agent.

Health Effects: Can cause allergic reactions. Many avoid allergic reaction by laundering garment before wearing. Generally harmless when used for intended purposes.

Synonyms: CAS: 589-41-3 ❖ ETHYL-N-HYDROXYCARBAMATE ❖ HYDROXYCARBAMIC ACID ETHYL ESTER ❖ N-HYDROXYURETHANE

HYDROXY ETHYL CELLULOSE

Products: In papers and textiles.

Use: As a thickening and suspending agent for products.

Health Effects: Can cause allergic reactions. Generally harmless when used for intended purposes.

Synonyms: CELLOSIZE

HYDROXYLAMINE SULFATE

Products: In paints, varnishes, and rustproofers.

Use: Various.

Health Effects: A corrosive irritant to skin, eyes, nose, and throat.

Synonyms: CAS: 10039-54-0 ❖ HYDROXYLAMMONIUM SULFATE ❖ HYDROXYLAMINE NEUTRAL SULFATE ❖ HYDROXYLAMINE SULFATE ❖ HYDROXYLAMMONIUM SULFATE

HYDROXYLATED LECITHIN

Products: In bakery products, beet sugar, beverages (dry mix), margarine, and yeast.

Use: As a clouding agent and emulsifier.

Health Effects: FDA approves use at moderate levels to accomplish the desired results.

Synonyms: None found.

2-HYDROXY-3-METHYL-2-CYCLOPENTEN-1-ONE

Products: In bakery products, beverages (nonalcoholic), chewing gum, confections, gelatin desserts, ice cream, puddings, and syrups.

Use: As a flavoring agent, food additive, and fragrancing agent.

Health Effects: In large, concentrated amounts, it is moderately toxic by swallowing. A human mutagen (changes inherited characteristics).

Synonyms: CAS: 80-71-7 ✽ CORYLON ✽ CORYLONE ✽ CYCLOTEN ✽ MAPLE LACTONE ✽ 3-METHYLCYCLOPENTANE-1,2-DIONE ✽ METHYL CYCLOPENTENOLONE

p-HYDROXYPROPYL BENZOATE

Products: In baked goods, beverages, food colors, milk, sausage (dry), and wine.

Use: A food additive, preservative, and antimicrobial

Health Effects: Mildly toxic by swallowing large, concentrated amounts. An allergen. FDA approves use within stated limitations.

Synonyms: CAS: 94-13-3 ✽ ASEPTOFORM P ✽ BETACIDE P ✽ BONOMOLD OP ✽ 4-HYDROXYBENZOIC ACID PROPYL ESTER ✽ p-HYDROXYBENZOIC ACID PROPYL ESTER ✽ NIPASOL ✽ p-OXYBENZOESAUREPROPYLESTER ✽ PARABEN ✽ PARASEPT ✽ PASEPTOL ✽ PRESERVAL P ✽ PROPYL p-HYDROXYBENZOATE ✽ n-PROPYL p-HYDROXYBENZOATE ✽ PROPYLPARABEN ✽ PROPYLPARASEPT ✽ PROTABEN P ✽ TEGOSEPT P

HYDROXYPROPYL CELLULOSE

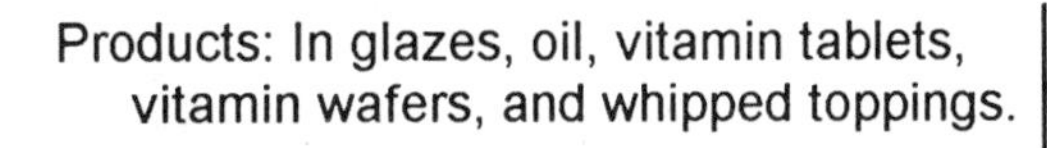

Products: In glazes, oil, vitamin tablets, vitamin wafers, and whipped toppings.

Use: An emulsifier, film former, stabilizer, suspending agent, and thickening agent.

Health Effects: Slightly toxic by swallowing pure, gross amounts.

Synonyms: CAS: 9004-64-2 ✽ HYDROXYPROPYL ETHER of CELLULOSE ✽ KLUCEL

HYDROXYPROPYL METHYLCELLULOSE

Products: In bakery goods, breading, dressings, and salad dressing mix.

Use: As emulsifier, stabilizer, and thickening agent. Also used as a thickener in paint-stripping preparations.

Health Effects: Harmless when used for intended purposes.

Synonyms: CAS: 9004-65-3 ✽ METHOCEL HG

8-HYDROXYQUINOLINE SULFATE

Products: In antiseptics, personal care products, toiletries, and fungicides.

Use: As bactericide, antiperspirant, deodorant, and fungus destroyer.

Health Effects: Poison by swallowing.

Synonyms: CAS: 134-31-6 ✽ OXINE SULFATE ✽ OXYQUINOLINE SULFATE

HYDROXYTRIPHENYLSTANNANE

Products: For beef, carrots, goat, horse, lamb, peanut hulls, peanuts, pecans, pork, potatoes, and sugar beet roots.

Use: To kill fungus on food products.

Health Effects: In large amounts it is a poison by swallowing the concentrated substance. A severe eye irritant.

Synonyms: CAS: 76-87-9 ❊ DU-TER ❊ FENOLOVO ❊ FENTIN HYDROXIDE ❊ HAITIN ❊ HYDROXYTRIPHENYLTIN ❊ SUZU H ❊ TPTH ❊ TRIPHENYLTIN OXIDE ❊ TUBOTIN

HYPOCHLOROUS ACID

Products: In textile bleaches, water purification, pool cleaners, and antiseptics.

Use: In bleaching, purifying, cleaning, and germ killing.

Health Effects: An irritant to skin and eyes.

Synonyms: CAS: 7790-92-3 ❊ HYPOCHLORITE SOLUTION ❊ CALCIUM HYPOCHLORITE ❊ SODIUM HYPOCHLORITE

HYPOPHOSPHORIC ACID

Products: Baking powder crystals.

Use: In baking and cooking.

Health Effects: Harmless when used for intended purpose.

Synonyms: CAS: 7803-60-3 ❊ SODIUM SALT ❊ PHOSPHINIC ACID ❊ ORTHOPHOSPHOROUS ACID

I

"The Environmental Protection Agency is adding 38 contaminants--including mercury, PCB's and a range of farm chemicals--to the current list of 24 for which communities must monitor their water four times a year."

ICELAND MOSS

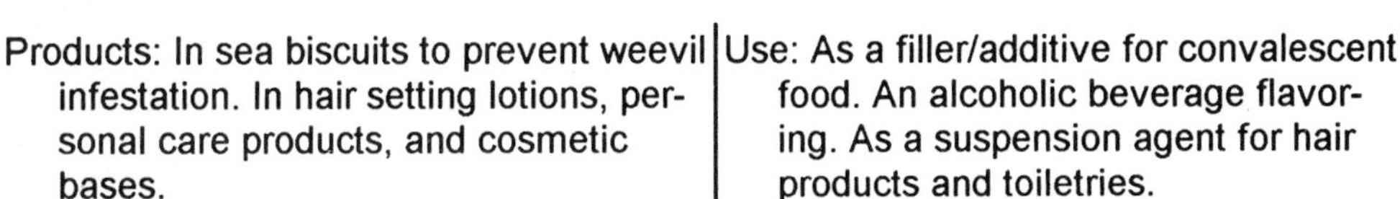

Products: In sea biscuits to prevent weevil infestation. In hair setting lotions, personal care products, and cosmetic bases.

Use: As a filler/additive for convalescent food. An alcoholic beverage flavoring. As a suspension agent for hair products and toiletries.

Health Effects: Harmless when used for intended purposes.

Synonyms: SCANDINAVIAN MOSS ✲ ICELAND LICHEN

ICHTHAMMOL

Products: Pharmaceutical products such as skin ointments, preparations, cosmetics, toiletries, and special dermatological soaps.

Use: An antiseptic, emollient (skin softener), and demulcent (relieves skin irritation).

Health Effects: Harmless when used for intended purposes.

Synonyms: AMMONIUM ICHTHOSULFONATE ✲ BITUMOL ✲ BITUMINOL ✲ ICHTHIUM ✲ ICHTOPUR ✲ LITHOL ✲ PETROSULPHOL ✲ TUMENOL

INDANTHRONE

Products: In fabric dyes; pigments for paints and enamels.

Use: As a coloring agent.

Health Effects: Harmless when used for intended purposes.

Synonyms: CAS: 81-77-6 ✲ INDANTHRENE BLUE R ✲ 6,15-DIHYDRO-5,14,18-ANTHRAZINETETRONE

INDIAN RED

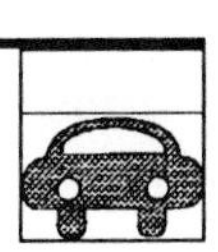

Products: In paint, rubber, plastics, and car polishes.

Use: As a coloring agent and rubbing polish.

Health Effects: Harmless when used for intended purposes.

Synonyms: IRON SAFFRON ✽ MAROON PIGMENT

INDOLE

Products: In perfumes for consumer toiletries and personal care products. (Unpleasant odor in high concentrations, but diluted solutions are pleasant).

Use: For aromas and scents.

Health Effects: Moderately toxic by swallowing and skin contact. FDA approves use at moderate levels to accomplish the intended effect.

Synonyms: CAS: 120-72-9 ✽ 1-AZAINDENE ✽ 1-BENZAZOLE ✽ BENZOPYRROLE ✽ 2,3-BENZOPYRROLE ✽ KETOLE

INOSITOL

Products: Derived from vegetables, citrus fruits, cereal grains, liver, kidney, and other meats. In dietary supplement of vitamin B. In medications.

Use: For nutritional supplementation.

Health Effects: GRAS (generally regarded as safe). Harmless when used for intended purposes.

Synonyms: CAS: 87-89-8 ✽ cis-1,2,3,5-trans-4,6-CYCLOHEXANEHEXOL ✽ i-INOSITOL ✽ meso-INOSITOL

INVERT SOAPS

Products: In detergents and soaps.

Use: For disinfecting and cleansing.

Health Effects: Harmless when used for intended purposes.

Synonyms: None found.

INVERT SUGAR

Products: In candy, icings, soft drinks, brewing, and humectant (moisturizer which prevents drying). A 50-50 mixture of two sugars, dextrose and fructose. It is sweeter and more soluble than table sugar.

Use: As a sweetener (nutritive) and moisturizing additive.

Health Effects: GRAS (generally regarded as safe). Found in low quality foods, causes tooth decay and should be avoided.

Synonyms: CAS: 8013-17-0 ✽ INVERT SUGAR SYRUP

IODINE

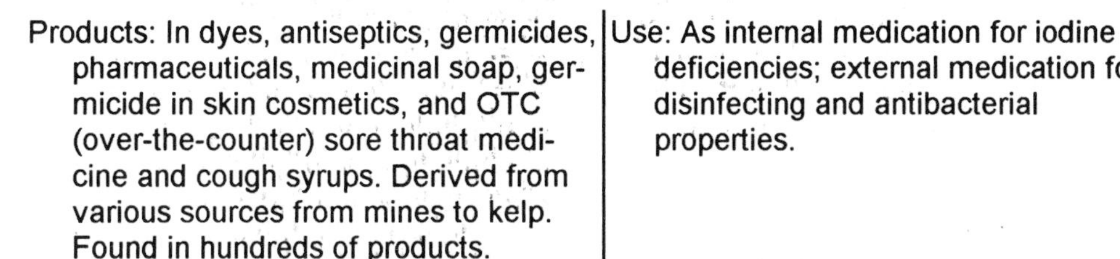

Products: In dyes, antiseptics, germicides, pharmaceuticals, medicinal soap, germicide in skin cosmetics, and OTC (over-the-counter) sore throat medicine and cough syrups. Derived from various sources from mines to kelp. Found in hundreds of products.

Use: As internal medication for iodine deficiencies; external medication for disinfecting and antibacterial properties.

Health Effects: Toxic by swallowing large amounts and by breathing fumes. Strong irritant to eyes and skin. Could cause allergic reaction.

Synonyms: CAS: 7553-56-2 ❃ JOD ❃ IODE ❃ IODINE CRYSTALS

IODINE TINCTURE

Products: In solution of iodine and potassium iodide in alcohol. An antiseptic.

Use: As a germicide.

Health Effects: Toxic by swallowing. Should not be used on open cuts because of toxicity.

Synonyms: None found.

IODOFORM

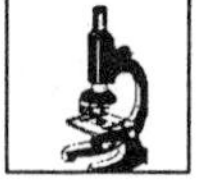

Products: In antiseptics and disinfectants.

Use: In medications.

Health Effects: A poison by swallowing. Moderately toxic by breathing and skin contact.

Synonyms: CAS: 75-47-8 ❃ TRIODOMETHANE

IPECAC

Products: A medical emetic (causes vomiting) and expectorant.

Use: To induce vomiting after swallowing toxic products.

Health Effects: Toxic. Label directions must be followed carefully.

Synonyms: CEPHALIS IPECACUANHA

IRON

Products: In enriched baked goods, cereal products, flour, pasta, and milk.

Use: As a dietary supplement or nutritional supplement.

Health Effects: Very beneficial in appropriate amounts. In gross, concentrated amounts it is a possible carcinogen (may cause cancer) that caused tumor growth in animals. Iron is potentially toxic in all forms and by all routes of exposure. Iron supplements are the leading cause of accidental poisoning deaths for children under six, who often mistake the pills for candy. Chronic exposure to excess levels of iron can result in pathological deposits of iron in the body tissues, the symptoms of which are fibrosis of the pancreas, diabetes mellitus, and liver cirrhosis.

Synonyms: CAS: 7439-89-6 ❊ ANCOR EN 80/150 ❊ ARMCO IRON ❊ CARBONYL IRON ❊ IRON, CARBONYL ❊ IRON, ELECTROLYTIC ❊ IRON, ELEMENTAL ❊ IRON, REDUCED

IRON BLUE

Products: In paints, printing inks, artists colors, cosmetic eye shadow, laundry blue, dyes, and finishes.

Use: As a pigment or coloring agent.

Health Effects: Harmless when used for intended purposes.

Synonyms: None found.

IRON OXIDE

Products: In pigments, cat food, dog food, and packaging materials.

Use: As a color additive, cosmetic color, and constituent of paperboard.

Health Effects: FDA approves use at limited levels to accomplish the desired results.

Synonyms: CAS: 1309-37-1 ❊ BAUXITE RESIDUE ❊ BLACK OXIDE of IRON ❊ BLENDED RED OXIDES of IRON ❊ BURNTISLAND RED ❊ BURNT SIENNA ❊ BURNT UMBER ❊ CALCOTONE RED ❊ COLCOTHAR ❊ COLLOIDAL FERRIC OXIDE ❊ FERRIC OXIDE ❊ INDIAN RED ❊ IRON(III) OXIDE ❊ IRON OXIDE RED ❊ IRON SESQUIOXIDE ❊ JEWELER'S ROUGE ❊ MARS BROWN ❊ MARS RED ❊ NATURAL IRON OXIDES ❊ NATURAL RED OXIDE ❊ OCHRE ❊ PRUSSIAN BROWN ❊ RED IRON OXIDE ❊ RED OCHRE ❊ ROUGE ❊ RUBIGO ❊ SIENNA ❊ SYNTHETIC IRON OXIDE ❊ VENETIAN RED ❊ VITRIOL RED ❊ YELLOW OXIDE of IRON

IRON OXIDE RED

Products: In marine paints, metal primers, polishing compounds, theatrical rouge, blush, cosmetics, and grease paints.

Use: As a pigment or coloring for various products.

Health Effects: Harmless when used for intended purposes.

Synonyms: CAS: 1332-37-2 ❊ BURNT SIENNA ❊ INDIAN RED ❊ RED IRON OXIDE ❊ RED OXIDE ❊ ROUGE ❊ TURKEY RED

ISOAMYL ACETATE

Products: In beverages, candy, and ice cream.

Use: Flavoring agent and odorants.

Health Effects: Mildly toxic by swallowing excessive amounts of pure substance and by breathing pure chemical. Exposure can cause headache, fatigue, pulmonary irritation, and serious toxicity effects. FDA approves use within stated limits.

Synonyms: CAS: 123-92-2 ✲ ACETIC ACID, ISOPENTYL ESTER ✲ BANANA OIL ✲ ISOAMYL ETHANOATE ✲ ISOPENTYL ACETATE ✲ ISOPENTYL ALCOHOL ACETATE ✲ 3-METHYLBUTYL ACETATE ✲ 3-METHYL-1-BUTYL ACETATE ✲ 3-METHYLBUTYL ETHANOATE ✲ PEAR OIL

ISOAMYL BENZOATE

Products: In perfumes, colognes, shaving lotions, cosmetics, and flavorings.

Use: A fruity odorant.

Health Effects: Mildly toxic by swallowing. A skin irritant.

Synonyms: CAS: 94-46-2 ✲ AMYL BENZOATE ✲ BENZOIC ACID, 1-(3-METHYL)BUTYL ESTER ✲ FEMA No. 2058 ✲ ISOPENTYL BENZOATE ✲ 1-(3-METHYL)BUTYL BENZOATE

ISOAMYL BENZYL ETHER

Products: Perfume ingredient in soaps.

Use: As a fragrancing agent.

Health Effects: A skin irritant. Could cause allergic reaction.

Synonyms: CAS: 122-73-6 ✲ BENZYL ISOAMYL ETHER ✲ BENZYL ISOPENTYL ETHER

ISOAMYL BUTYRATE

Products: In baked goods, dessert gels, and puddings.

Use: A fruity-odored flavoring agent.

Health Effects: Harmless when used for intended purposes. FDA approves use at levels to accomplish the intended effect.

Synonyms: CAS: 106-27-4 ✲ AMYL BUTYRATE

ISOAMYL FORMATE

Products: In candy, dessert gels, ice cream, and puddings.

Use: An artificial fruit-essence flavoring.

Health Effects: In large amounts it is moderately toxic by swallowing concentrated chemical. A skin irritant. Concentrated material is very irritating and can cause narcosis (numbing). The symptoms are usually brief, but it is possible upon severe or prolonged exposure to have serious consequences. FDA approves use at moderate levels to accomplish the desired results.

Synonyms: CAS: 110-45-2 ✲ FORMIC ACID, ISOPENTYL ESTER ✲ ISOAMYL METHANOATE ✲ ISOPENTYL ALCOHOL, FORMATE ✲ ISOPENTYL FORMATE ✲ 3-METHYLBUTYL FORMATE

ISOAMYL HEXANOATE

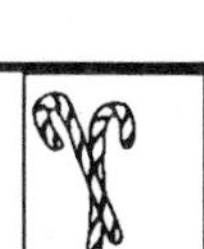

Products: In candy, desserts, and ice

Use: A fruity-odored flavoring agent.

cream.

Health Effects: A mild skin irritant in concentrated form.

Synonyms: CAS: 2198-61-0 ✲ AMYL HEXANOATE ✲ ISOAMYL CAPROATE

ISOAMYL SALICYLATE

Products: A floral (orchid) odor. In soaps, skin products, toiletries, and shaving products. An ingredient that affects the smell of final product.

Use: As a fragrancing or scenting agent.

Health Effects: Could cause allergic reactions in susceptible individuals.

Synonyms: CAS: 87-20-7 ✲ AMYL SALICYLATE ✲ ISOAMYL o-HYDROXYBENZOATE ✲ SALICYLIC ACID, ISOPENTYL ESTER

ISOAMYL VALERATE

Products: Apple and fruit flavoring in beverages, ice creams, gelatins, spreads, and alcoholic liqueurs.

Use: A fruity-essence seasoning for consumables.

Health Effects: Harmless when used for intended purposes.

Synonyms: APPLE ESSENCE ✲ APPLE OIL ✲ ISOAMYL ISOVALERATE ✲ AMYL VALERIANATE ✲ AMYL VALERATE

ISOBORNYL ACETATE

Products: In toilet waters, bubble bath, bath oils, antiseptics, air fresheners, soaps, and food flavorings.

Use: A pine-needle odorant and essence.

Health Effects: Could cause allergic reaction.

Synonyms: None found.

ISOBORNYL SALICYLATE

Products: In perfumes, toiletries, cosmetics, suntan preparations, oils, lotions, mousses, and gels.

Use: As a fixative and filter in preparations.

Health Effects: A possible allergen in susceptible individuals.

Synonyms: None found.

ISOBUTYL ACETATE

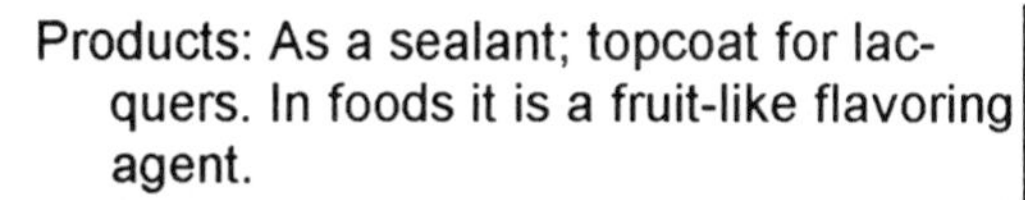

Products: As a sealant; topcoat for lacquers. In foods it is a fruit-like flavoring agent.

Use: As solvent, perfume, and seasoning agent.

Health Effects: In excessive amounts it is mildly toxic by swallowing and by breathing fumes. A skin and eye irritant. Approved by the FDA for use at moderate levels to accomplish the desired results.

Synonyms: CAS: 110-19-0 ❊ ACETIC ACID, ISOBUTYL ESTER ❊ ACETIC ACID-2-METHYLPROPYL ESTER ❊ 2-METHYLPROPYL ACETATE ❊ 2-METHYL-1-PROPYL ACETATE ❊ β-METHYLPROPYL ETHANOATE

ISOBUTYL ALCOHOL

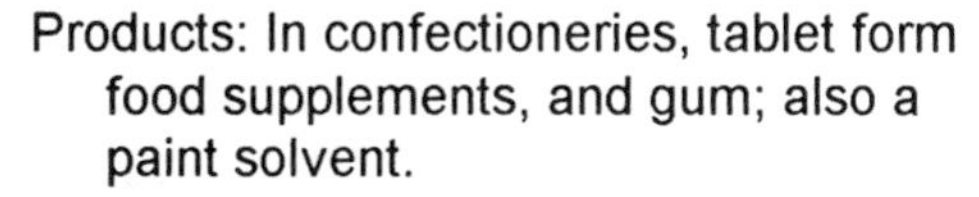

Products: In confectioneries, tablet form food supplements, and gum; also a paint solvent.

Use: A fruit-flavored food concentrate. Paint remover.

Health Effects: Pure chemical is moderately toxic by swallowing large amounts and by skin contact. Mildly toxic by breathing fumes. A severe skin and eye irritant. A possible carcinogen.

Synonyms: CAS: 78-83-1 ❊ FERMENTATION BUTYL ALCOHOL ❊ 1-HYDROXYMETHYLPROPANE ❊ ISOBUTANOL ❊ ISOPROPYLCARBINOL ❊ 2-METHYL PROPANOL ❊ 2-METHYL-1-PROPANOL ❊ 2-METHYLPROPAN-1-OL ❊ 2-METHYLPROPYL ALCOHOL

ISOBUTYL-p-AMINOBENZOATE

Products: In topical anesthetic and suntan oils, creams, lotions, gels, mousses and preparations.

Use: As medication and sunscreen protectorants.

Health Effects: Harmless when used for intended purposes.

Synonyms: CAS: 94-14-4 ❊ p-AMINOBENZOIC ACID ISOBUTYL ESTER ❊ (2-METHYL)-p-AMINOBENZOATE

ISOBUTYLENE-ISOPRENE COPOLYMER

Products: In chewing gum, gumballs, sealants, coatings, finishings, cements, and adhesives.

Use: Masticatory substance in chewing gum base, insulating, sealing, and adherents.

Health Effects: Harmless when used for intended purposes. FDA approves use in amounts to accomplish the intended effect.

Synonyms: BUTYL RUBBER

ISOBUTYL STEARATE

Products: In coatings, polishes, inks, cosmetics, and soaps.

Use: As a waterproofer and as makeup cream and blush.

Health Effects: Harmless when used for intended purposes.

Synonyms: CAS: 646-13-9

ISOBUTYRIC ACID

Products: In varnish, flavor, and perfume bases.

Use: As disinfecting agent, solvent, and seasoning.

Health Effects: A poison by swallowing. Moderately toxic by skin contact. A corrosive irritant to the eyes, skin, nose, and throats.

Synonyms: CAS: 79-31-2 ✲ DIMETHYLACETIC ACID ✲ ISOPROPYLFORMIC ACID ✲ α-METHYLPROPIONIC ACID ✲ 2-METHYLPROPANOIC ACID ✲ 2-METHYLPROPIONIC ACID

ISODECYL CHLORIDE

Products: In oils, fats, greases, pharmaceuticals, and detergents.

Use: For cleaning compounds and solvents.

Health Effects: Harmless when used for intended purposes.

Synonyms: None found.

ISOEUGENOL

Products: In room deodorants, toiletries, colognes, hair tonics, and hand creams.

Use: Fragrance variously described as spice-clove, floral-carnation, or vanillin.

Health Effects: Moderately toxic by swallowing. A human mutagen (changes inherited characteristics).

Synonyms: CAS: 97-54-1 ✲ 1-HYDROXY-2-METHOXY-4-PROPENYLBENZENE ✲ 4-HYDROXY-3-METHOXY-1-PROPENYLBENZENE ✲ 2-METHOXY-4-PROPENYLPHENOL ✲ 4-PROPENYLGUAIACOL

ISOPARAFFINIC PETROLEUM HYDROCARBONS, SYNTHETIC

Products: For eggs, fruits, pickles, vegetables, vinegar, and wine.

Use: As a coating agent; insecticide formulations component.

Health Effects: FDA approves use at moderate levels to accomplish the intended effects.

Synonyms: SYNTHETIC ISOPARAFFINIC PETROLEUM HYDROCARBONS

ISOPROPANOLAMINE

Products: In drycleaning soaps, wax removers, and cosmetics.

Use: An emulsifying agent (stabilizes and maintains mixes to aid in suspension of oily liquid).

Health Effects: Harmless when used for intended purposes.

Synonyms: CAS: 78-96-6 ❊ 2-HYDROXYPROPYLAMINE ❊ 1-AMINO-2-PROPANOL ❊ MIPA

ISOPROPYL ACETATE

Products: In paints, lacquers, inks, and perfumes.

Use: As solvent and aromatic.

Health Effects: Moderately toxic by swallowing. Mildly toxic by breathing. Harmful effects on the human body by breathing vapors. Narcotic in high concentration. Chronic exposure can cause liver damage.

Synonyms: CAS: 108-21-4 ❊ ACETIC ACID ISOPROPYL ESTER ❊ 2-ACETOXYPROPANE ❊ 2-PROPYL ACETATE

N-ISOPROPYLACRYLAMIDE

Products: In textiles, paper, adhesives, detergents, and cosmetics as a binder (used for holding substances together).

Use: As a processing aid.

Health Effects: Could cause allergic reaction. Harmless when used for intended purposes.

Synonyms: NIPAM

ISOPROPYL ALCOHOL

Products: In lotions, antifreeze, shellac solvent, quick-drying inks, toiletries, body lotions, after-shave lotions, hair preparations, and cosmetics.

Use: As a deicer, preservative, solvent, and dehydrating (drying) agent.

Health Effects: Can cause corneal burns and irritation. Moderately toxic to humans. Mildly toxic by skin contact. Effects by swallowing or breathing: flushing, pulse rate decrease, blood pressure lowering, anesthesia, narcosis, headache, dizziness, mental depression, hallucinations, distorted perceptions, dyspnea (shortness of breath), respiratory depression, nausea, vomiting, and coma. A mutagen (changes inherited characteristics).

Synonyms: CAS: 67-63-0 ❊ DIMETHYLCARBINOL ❊ ISOHOL ❊ ISOPROPANOL ❊ LUTOSOL ❊ PETROHOL ❊ PROPAN-2-OL ❊ 2-PROPANOL ❊ sec-PROPYL ALCOHOL ❊ i-PROPYLALKOHOL ❊ SPECTRAR

ISOPROPYLAMINE

Products: In pharmaceuticals, dyes, and bactericides.

Use: As a solvent, insecticide, germicide, among others.

Health Effects: Poison by skin contact in concentrated form. Moderately toxic by swallowing. Mildly toxic by breathing. A severe skin and eye irritant. A narcotic in high concentrations. Could cause allergic response.

Synonyms: CAS: 75-31-0 ✲ 2-AMINOPROPANE ✲ MONOISPROPYLAMINE ✲ 2-PROPANAMINE

ISOPROPYL CITRATE

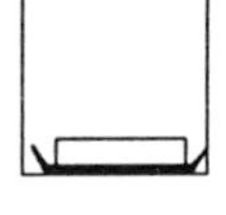

Products: In oleomargarine and vegetable oils.

Use: A preservative, sequestrant (affects the final products appearance, flavor, or texture).

Health Effects: FDA states GRAS when used within limitations.

Synonyms: None found.

ISOPROPYL ETHER

Products: Paint remover, varnish remover, spot remover, and rubber cements.

Use: In solvents and adhesives.

Health Effects: Mildly toxic by swallowing, breathing, and skin contact. A skin irritant.

Synonyms: CAS: 108-20-3 ✲ DIISOPROPYL ETHER ✲ DIISOPROPYL OXIDE ✲ 2-ISOPROPOXYPROPANE

ISOPROPYL MYRISTATE

Products: In cosmetic creams, toiletries, suntan lotions, and topical skin medications.

Use: As a preservative (it prevents rancidity). It is not easily soluble in water and maintains long lasting effects. It improves absorption through skin.

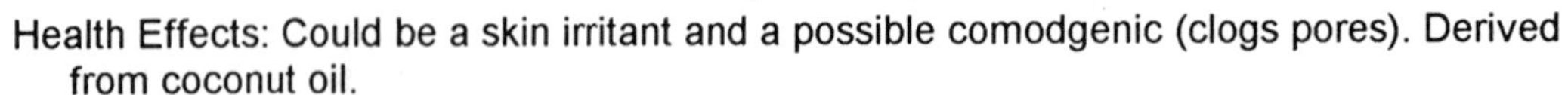

Health Effects: Could be a skin irritant and a possible comodgenic (clogs pores). Derived from coconut oil.

Synonyms: CAS: 110-27-0 ✲ ISOMYST ✲ KESSCOMIR ✲ TETRADECANOIC ACID, ISOPROPYL

ISOPROPYL PALMITATE

Products: In cosmetic lotions, creams, deodorants, and skin care products.

Use: As an emollient (skin softener), and emulsifier (stabilizes and suspends oils in mixture).

Health Effects: Could cause allergic reaction.

Synonyms: CAS: 142-91-6 ❊ CRODAMOL IPP ❊ DELTYL ❊ HEXANDECANOIC ACID ❊ ISOPROPYL ESTER ❊ ISOPAL ❊ PROPAL ❊ TEGESTER ISOPALM ❊ ISOPROPYL HEXADECANOATE ❊ DELTYL PRIME ❊ STEPAN D-70 ❊ UNIMATE IPP ❊ WICKENOL 111

ISOVALERIC ACID

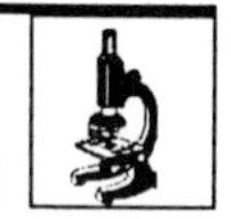

Products: In medicine, foods, and perfumes.

Use: To adjust and dilute flavorings and tastes in products.

Health Effects: In concentrated form, the chemical is a poison by skin contact. Moderately toxic by swallowing. A corrosive skin and eye irritant. FDA approves use at moderate levels to accomplish the intended effects.

Synonyms: CAS: 503-74-2 ❊ DELPHINIC ACID ❊ ISOPENTANOIC ACID ❊ ISOPROPYLACETIC ACID ❊ ISOVALERIANIC AICD ❊ 3-METHYLBUTANOIC ACID ❊ β-METHYLBUTYRIC ACID ❊ 3-METHYLBUTYRIC ACID

J

"People who get 180 mg of vitamin C have about 10 percent higher good type HDL cholesterol than those eating less."

Healthy Eating by Jean Carper

JAPAN WAX

Products: A substitute for beeswax. For candles, floor waxes, polishes, back plasters, ointments, furniture polish, and dental impression compounds.

Use: As polish, plasticizer, and coating.

Health Effects: Harmless when used for intended purposes.

Synonyms: JAPAN TALLOW ❊ SUMAC WAX

JASMINE OIL

Products: Perfume in colognes, toiletries, bath products, talcs, and hair products.

Use: An odorant.

Health Effects: Could cause allergic reaction.

Synonyms: JASMINE FLOWER ❊ SWEET JASMINE OIL

JOJOBA OIL

Products: Substitute for sperm oil for transmission lubricants. Substitute for carnauba wax and beeswax. For cosmetic preparations such as hair products, shampoos, conditioners, moisturizers, skin products, toiletries, softeners, hand lotions, and suntan products.

Use: As a lubricant and conditioner.

Health Effects: Possible allergic reactions in susceptible individuals.

Synonyms: DESERT OIL ❊ OIL OF JOJOBA

JUNIPER BERRY OIL

Products: In gin, root beer, wintergreen, ginger, and liqueurs.

Use: As a seasoning and flavoring for assorted food and beverage products.

Health Effects: Mildly toxic by swallowing large, pure amounts. A skin irritant. An allergen. Effects on the body by swallowing are severe kidney irritation similar to that caused by turpentine. GRAS (generally regarded as safe) when used at moderate levels to accomplish the intended effect.

Synonyms: CAS: 8012-91-7 ❋ OIL of JUNIPER BERRY ❋ WACHOLDERBEER OEL (GERMAN)

"Most Americans spend up to 90% of their lives indoors, and indoor air pollution is among the top environmental problems of the 1990's."

Environmental Protection Agency

KAOLIN

Products: For cements, fertilizers, cosmetic face powder, baby powder, bath powder, foundation facial makeup, paints, and antidiarrheal medications.

Use: In wine filtration, anticaking agent, clarifying agent, covering agent, filling agent, and assorted purposes.

Health Effects: Harmless when used for intended purposes.

Synonyms: CAS: 1332-58-7 ❖ ALTOWHITES ❖ BENTONE ❖ CONTINENTAL ❖ DIXIE ❖ EMATHLITE ❖ FITROL ❖ FITROL DESICCITE 25 ❖ GLOMAX ❖ HYDRITE ❖ KAOPAOUS ❖ KAOPHILLS-2 ❖ LANGFORD ❖ MCNAMEE ❖ PARCLAY ❖ PEERLESS ❖ SNOW TEX

KARAYA GUM

Products: In baked goods, candy (soft), frozen dairy desserts, milk products, and toppings.

Use: An emulsifier (maintains mixes to aid in suspension of oily liquids), stabilizer (maintains consistency), thickening agent, and adhesive.

Health Effects: In excessive amounts the pure chemical is mildly toxic by swallowing. A mild allergen. FDA approves use when used within limitations.

Synonyms: CAS: 9000-36-6 ❖ STERCULIA GUM ❖ INDIA TRAGACANTH ❖ KADAYA GUM

KELP

Products: An algae in chewing gum base, fertilizer, and dietary supplement.

Use: As filler, seasoning, or flavoring.

Health Effects: A large, coarse, seaweed found off the coast of California. Harmless when used for intended purposes. FDA states it is GRAS (generally regarded as safe).

Synonyms: MACROCYSTIS PYRIFERAE

KERATINASE

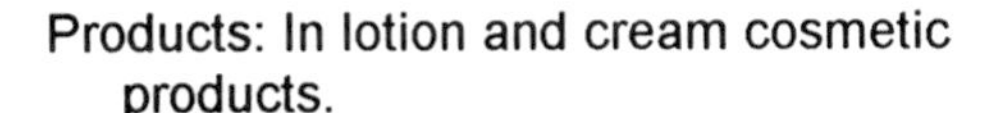

Products: In lotion and cream cosmetic products.

Use: As a depilatory (hair remover).

Health Effects: A potential skin irritant and allergen. Directions must be followed carefully on products containing this enzyme.

Synonyms: None found.

KEROSENE

Products: A fuel in lanterns, lamps, heaters, flares, and stoves. As a degreaser. As a solvent in cosmetics.

Use: As a combustible, solvent, and insect spray.

Health Effects: A possible carcinogen. A severe skin irritant. A mutagen (changes inherited characteristics). Effects on the body by swallowing are somnolence (sleepiness), hallucinations and distorted perceptions, coughing, nausea, vomiting, and fever. Aspiration of vomitus has caused pneumonia especially in children. Moderately explosive in the form of vapor when exposed to heat or flame.

Synonyms: CAS: 8008-20-6 ❊ COAL OIL ❊ DEOBASE ❊ STRAIGHT-RUN KEROSENE

KETONE

Products: In lacquers and paints.

Use: As a solvent and for textile processing.

Health Effects: Moderately toxic. A skin and severe eye irritant. Harmful effects on the body by breathing. Container directions must be carefully followed.

Synonyms: None found.

KOJIC ACID

Products: In insecticides, antifungals, and antimicrobial agents.

Use: Various.

Health Effects: Moderately toxic by swallowing.

Synonyms: CAS: 501-30-4 ❊ [5-HYDROXY-2-(HYDROXYMETHYL)-4-PYRONE]

L

Both the FDA and the World Health Organization have declared irradiation--bombing food with gamma rays to kill insects, bacteria and other organisms--to be safe.

LACQUER

Products: In metal, protective, or decorative coatings; paper products, textiles, plastics, furniture polish, and nail polish.

Use: As a plasticizer, coating, and polish.

Health Effects: Variable toxicity depending upon concentrations of products. They may cause allergic reactions. A dangerous fire hazard due to the highly flammable solvents commonly used. A severe explosion hazard in the form of vapor when exposed to flame.

Synonyms: NITROCELLULOSE-ALKYD LACQUERS ✲ NITROCELLULOSE ✲ VINYL RESINS ✲ ACRYLIC RESINS

LACTATED MONO-DIGLYCERIDES

Products: In margarine and oleomargarine.

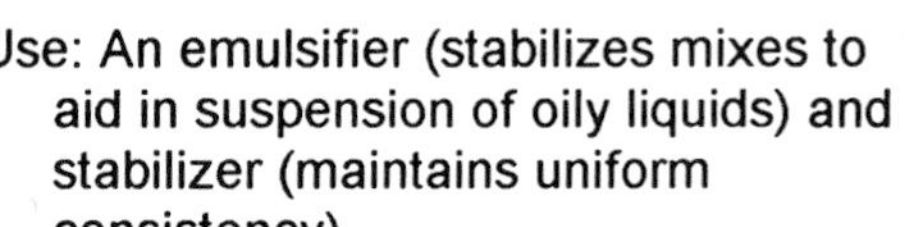

Use: An emulsifier (stabilizes mixes to aid in suspension of oily liquids) and stabilizer (maintains uniform consistency).

Health Effects: Harmless when used for intended purposes.

Synonyms: None found.

LACTIC ACID

Products: In cheese spreads, egg (dry powder), olives, poultry, salad dressing mix, and wine.

Use: As an acid, antimicrobial agent, curing agent, flavor enhancer, flavoring agent, pH control agent, pickling agent, solvent, and as a vehicle for other chemicals.

Health Effects: Moderately toxic by swallowing gross quantities of pure chemical. A severe skin and eye irritant. GRAS (generally regarded as safe) when used at moderate levels to accomplish the desired results.

Synonyms: CAS: 50-21-5 ✲ ACETONIC ACID ✲ ETHYLIDENELACTIC ACID ✲ 1-HYDROXYETHANECARBOXYLIC ACID ✲ 2-HYDROXYPROPANOIC ACID ✲ 2-HYDROXYPROPIONIC ACID ✲ α-HYDROXYPROPIONIC ACID ✲ KYSELINA MLECNA ✲ *dl*-LACTIC ACID ✲ MILCHSAURE ✲ MILK ACID ✲ ORDINARY LACTIC ACID ✲ RACEMIC LACTIC ACID

LACTIC ACID, MONOSODIUM SALT

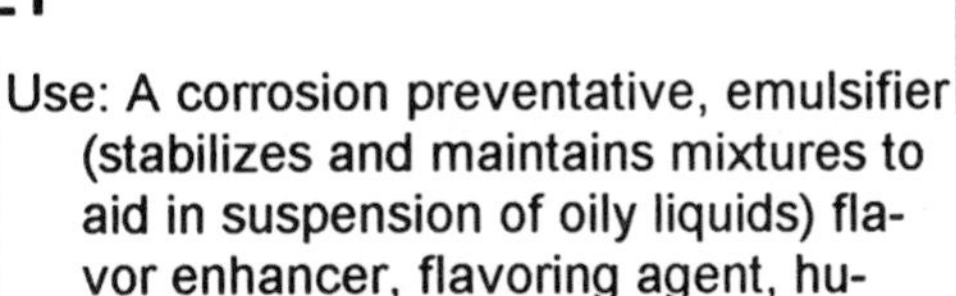

Products: In biscuits, fruits, meat products, nuts, sponge cake, Swiss roll, vegetables, water (bottled), and water (canned).

Use: A corrosion preventative, emulsifier (stabilizes and maintains mixtures to aid in suspension of oily liquids) flavor enhancer, flavoring agent, humectant (keeps product from drying out), and washing agent.

Health Effects: Chemical is an eye and skin irritant. GRAS (generally regarded as safe) when used within stated FDA limitations.

Synonyms: CAS: 72-17-3 ✲ 2-HYDROXYPROPANOIC ACID MONOSODIUM SALT ✲ LACOLIN ✲ LACTIC ACID SODIUM SALT ✲ PER-GLYCERIN ✲ SODIUM LACTATE

LACTOSE

Products: In infant foods, pharmaceutical bases, laxatives, baked goods, candies, margarine, and butter. Used in the manufacture of penicillin. Milk contains 5% lactose.

Use: As a nutrient; for baby food formulations.

Health Effects: Could cause allergic reaction. Harmless when used for intended purposes.

Synonyms: CAS: 63-42-3 ✲ LACTIN ✲ LACTOBIOSE ✲ *d*-LACTOSE ✲ MILK SUGAR ✲ SACCHARUM LACTIN

LACTYLATED FATTY ACID ESTERS of GLYCEROL and PROPYLENE GLYCOL

Products: In cake mixes, coffee whiteners, icings, and toppings.

Use: An emulsifier (stabilizes and maintains mixes to aid in suspension of oily liquids), stabilizer (used to keep uniform consistency), and whipping agent.

Health Effects: Harmless when used for intended purposes. Must conform to FDA specifications.

Synonyms: PROPYLENE GLYCOL LACTOSTEARATE

LACTYLIC ESTERS of FATTY ACIDS

Products: In baked goods, bakery mixes, frozen desserts, fruit juices (dehydrated), fruits (dehydrated), milk or cream substitutes for beverage coffee, pancake mixes, pudding mixes, rice (precooked instant), shortening (liquid), vegetable juices (dehydrated), and vegetables (dehydrated).

Use: An emulsifier (stabilizes and maintains mixes to aid in suspension of oily liquids).

Health Effects: Harmless when used for intended purposes. FDA approves use at a moderate level to accomplish the desired results.

Synonyms: None found.

LAMPBLACK

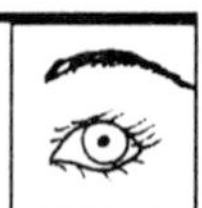

Products: Pigment for cements, inks, crayons, polishes, carbon paper, soap, matches, and cosmetic eye-liner pencils.

Use: For coloring, coating, lubricating, and polishing.

Health Effects: Harmless when used for intended purposes while taking normal precautions.

Synonyms: CARBONACEOUS MATERIAL ❖ MICROCRYSTALLINE CARBON

LANOLIN

Products: In ointments, leather polishes, face creams, facial tissue, hair dressing products, cosmetics, lipstick, mascara, rouge, eye shadow, and suntan preparations. Lanolin anhydrous is used in chewing gum as a chewing gum base.

Use: Primarily a moisturizer.

Health Effects: Derived from sheep oil glands. A frequent cause of allergic reactions to the skin.

Synonyms: WOOL FAT

LARD

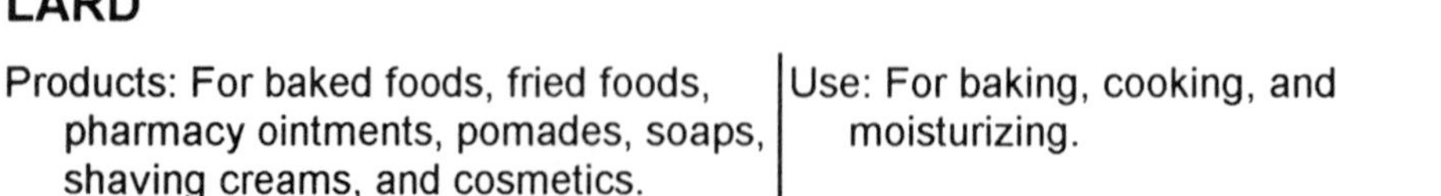

Products: For baked foods, fried foods, pharmacy ointments, pomades, soaps, shaving creams, and cosmetics.

Use: For baking, cooking, and moisturizing.

Health Effects: High in saturated fats. Otherwise harmless when used for intended purposes.
Synonyms: HOG FAT ✲ STEARIN ✲ PALMITIN ✲ OLEIN ✲ PORK FAT ✲ PORK OIL

LARD (UNHYDROGENATED)

Products: In cake mixes.

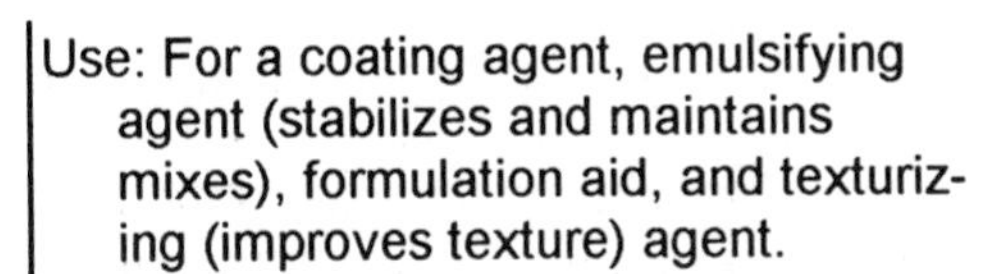

Use: For a coating agent, emulsifying agent (stabilizes and maintains mixes), formulation aid, and texturizing (improves texture) agent.

Health Effects: Harmless when used for intended purposes. GRAS (generally regarded as safe).
Synonyms: BLEACHED LARD ✲ BLEACHED-DEODORIZED LARD

LATEX

Products: In gloves, toys, adhesives, medical equipment, undergarments, and personal products.

Use: For flexibility, strengthening, and waterproofing, of consumer items.

Health Effects: Contact could cause an allergic reaction to skin. Causes rashes, irritation, flakiness, and itching. Occurs naturally in some species of trees and plants or can be made synthetically.
Synonyms: RUBBER HYDROCARBON ✲ STYRENE-BUTADIENE COPOLYMER ✲ ACRYLATE RESINS ✲ POLYVINYL ACETATE ✲ NATURAL LATEX

LAURIC ACID

Products: In coconut oil, laurel oil, vegetable fats, soaps, shampoos, detergents, and flavorings in foods.

Use: As a lubricant, soap, foam, or bubble agent. As a flavor enhancer in foods.

Health Effects: In large amounts the chemical is mildly toxic by swallowing.
Synonyms: CAS: 143-07-7 ✲ DODECANOIC ACID ✲ DODECOIC ACID ✲ DUODECYLIC ACID ✲ HYDROFOL ACID ✲ HYSTRENE ✲ LAUROSTEARIC ACID ✲ NEO-FAT ✲ NINOL EXTRA ✲ 1-UNDECANECARBOXYLIC ACID ✲ WECOLINE

LAUROYL PEROXIDE

Products: In fats, oils, waxes, cleaners, polishes, furniture, and flooring.

Use: As a bleaching or drying agent.

Health Effects: Toxic by swallowing and breathing. Corrosive to eyes, nose, and throats.
Synonyms: CAS: 105-74-8 ✲ DODECANOLY PEROXIDE ✲ ALPEROX C ✲ LAUROX ✲ LAURYDOL

LAURYL SULFOACETATE

Products: In toothpaste, tooth powder, liquid dentifrices, foaming bath products, shampoos, and synthetic detergents.

Use: As a biodegradable, detergent, wetting, scouring, emulsifying, dispersing, and foaming agent.

Health Effects: Harmless when used for intended purposes.

Synonyms: LANTHANOL ✲ SODIUM LAURYL SULFOACETATE

LAVENDER OIL

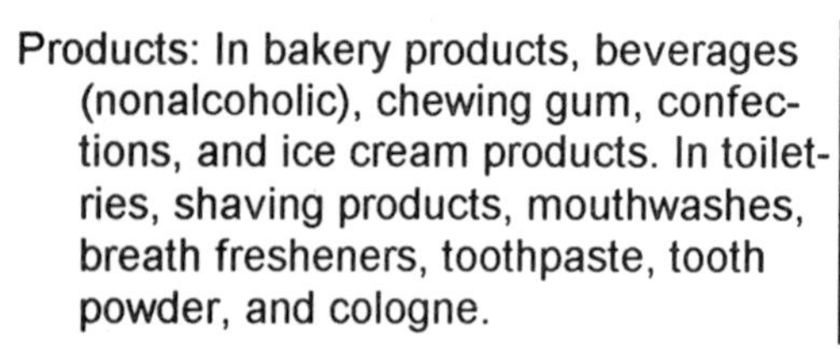

Products: In bakery products, beverages (nonalcoholic), chewing gum, confections, and ice cream products. In toiletries, shaving products, mouthwashes, breath fresheners, toothpaste, tooth powder, and cologne.

Use: As a seasoning or fragrancing agent. An ingredient to affect the taste or smell of final product.

Health Effects: In an excessive amount it is mildly toxic by swallowing pure chemical. A skin irritant. GRAS (generally regarded as safe) when used at moderate levels to accomplish the desired results.

Synonyms: CAS: 8000-28-0 ✲ LAVENDEL OEL (GERMAN) ✲ OIL of LAVENDER

LEAD

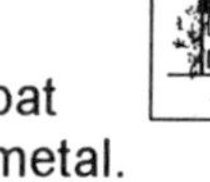

Products: Paint pigment, solder, glazes on ceramic dishes and bowls. Prohibited from interstate commerce since the middle 70's. It is still manufactured and used locally. In marine paints and bridge paints. Lead has been found in wines, possibly from the foil used on bottles. Pencil "lead" is not lead at all, rather it is a mixture of graphite and clay.

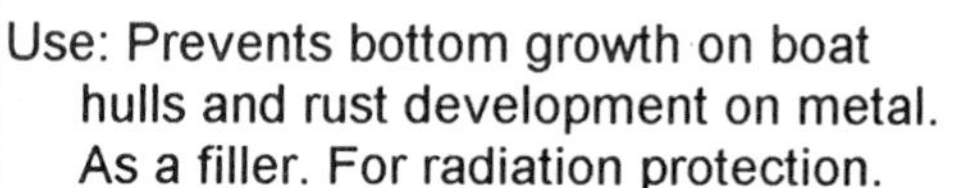

Use: Prevents bottom growth on boat hulls and rust development on metal. As a filler. For radiation protection.

Health Effects: Suspected carcinogen. Poison by swallowing. Effects on the human body by swallowing and breathing are: loss of appetite, anemia, malaise, insomnia, headache, irritability, muscle pains, joint pains, tremors, hallucinations, distorted perceptions, muscle weakness, gastritis, and liver changes. The major organ systems affected are the nervous system, blood system, and kidneys.

Any amount of lead is unsafe for children, pregnant women and nursing mothers. Studies now suggests that blood levels of lead below 10 μg/dL can have the effect of lowering the IQ scores of children. Low levels of lead impair neurotransmission, immune system function, and may increase systolic blood pressure. Kidney damage can occur from exposure.

Severe toxicity can cause sterility, abortion, neonatal mortality, and morbidity (diseased state). A mutagen (changes inherited characteristics) reported. Very heavy intoxication can sometimes be detected by formation of a dark line on the gum margins, the so-called "lead line." For the general population, exposure to lead occurs from breathed air, dust of various types, and food and water with an approximate 50/50 division between breathing and swallowing routes. Lead occurs in water in either dissolved or particulate form. Commonly found in paint chips from older properties where children live. Children eat paint chips and develop chronic symptoms. Acidic foods (tomatoes, fruit juices, etc.) leach lead out of decorative glazes in chinaware and cause chronic symptoms in those who ingest food or drinks.

Synonyms: CAS: 7439-92-1 ❊ GLOVER ❊ LEAD FLAKE ❊ LEAD S2 ❊ OLOW (POLISH) ❊ OMAHA ❊ OMAHA & GRANT ❊ SI ❊ SO

LEAD ACETATE

Products: In astringents, dyes, and printing colors. For waterproofing, varnishes, antifouling paints, and hair dyes.

Use: Various.

Health Effects: A lead compound. Toxic by swallowing, breathing, and skin absorption.

Synonyms: CAS: 301-04-2 ❊ SUGAR OF LEAD

LEAD CARBONATE, BASIC

Products: Exterior paint pigments and ceramic glazes.

Use: As filler and coloring agent.

Health Effects: Toxic by breathing.

Synonyms: LEAD SUBCARBONATE ❊ WHITE LEAD ❊ LEAD FLAKE

LEAD NAPTHENATE

Products: Paint and varnish drier; wood preservative, and lubrication oil additive.

Use: As a coating drier, stabilizer, and friction reducer.

Health Effects: Toxic material. A known carcinogen (causes cancer). Absorbed by skin.

Synonyms: CAS: 61790-14-5

LEAD RESINATE

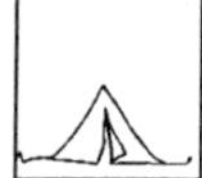

Products: For tents, umbrellas, clothing, camping equipment, and awnings.

Use: As a waterproofing agent.

Health Effects: Toxic material which can be absorbed by skin.

Synonyms: None found.

LECITHIN

Products: In baked goods, beverage powders, cocoa powder, fat (griddling), fillings, chocolate, meat products, oleomargarine, poultry, and shortening. In cosmetic products such as lotions, soaps, creams, eye makeup, and face makeup.

Use: An antioxidant (slows reaction with oxygen, slows spoiling), emulsifier (stabilizes and maintains mixes to aid in suspension of oily liquids).

Health Effects: GRAS (generally regarded as safe) when used within FDA limitations.

Synonyms: None found.

LEMONGRASS OIL

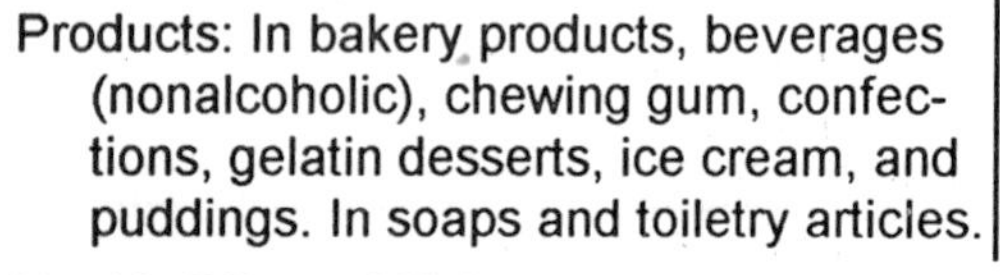

Products: In bakery products, beverages (nonalcoholic), chewing gum, confections, gelatin desserts, ice cream, and puddings. In soaps and toiletry articles.

Use: Flavoring agent in beverages and foods. Also in soaps and personal care items as odorant.

Health Effects: Mildly toxic by swallowing excessive amounts of pure chemical. A skin irritant in concentration. GRAS (generally regarded as safe) by FDA when used at moderate levels to accomplish the desired results.

Synonyms: CAS: 8007-02-1 ✲ GUATEMALA LEMONGRASS OIL ✲ MADAGASCAR LEMONGRASS OIL ✲ WEST INDIAN LEMONGRASS OIL

LEMON OIL

Products: In bakery products, beverages (nonalcoholic), cereals, chewing gum, condiments, confections, gelatin desserts, ice cream, meat, puddings, and syrups. In perfumes and air fresheners. Also in lemon oil furniture polish for wood.

Use: Polishing, flavoring and fragrancing agent.

Health Effects: Moderately toxic by swallowing large amounts of concentrated substance. A skin irritant. Could cause allergic reaction. GRAS (generally regarded as safe) when used at moderate level to accomplish the intended effects.

Synonyms: CAS: 8008-56-8 ✲ CEDRO OIL ✲ LEMON OIL, COLDPRESSED (FCC) ✲ LEMON OIL, EXPRESSED ✲ OIL of LEMON ✲ ZITRONEN OEL (GERMAN)

LEVULOSE

Products: In baked goods and beverages (low calorie).

Use: As a formulation aid, sweetener (nutritive), and processing oil.

Health Effects: Harmless when used for intended purposes.

Synonyms: CAS: 7660-25-5 ✲ FRUCTOSE (FCC) ✲ FRUIT SUGAR ✲ FRUTABS ✲ LAEVORAL ✲ LAEVOSAN ✲ LEVUGEN

LICORICE ROOT EXTRACT

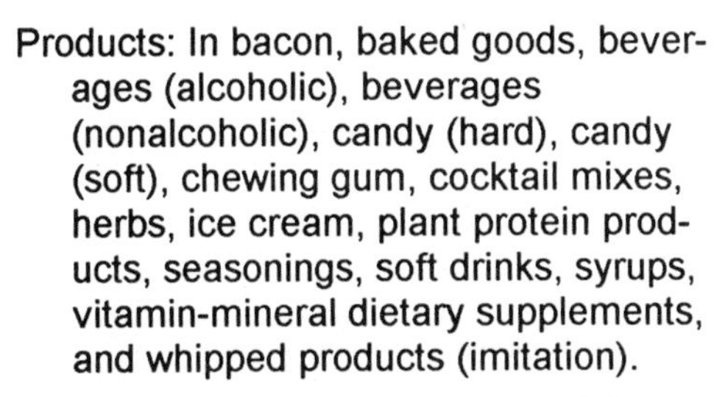

Products: In bacon, baked goods, beverages (alcoholic), beverages (nonalcoholic), candy (hard), candy (soft), chewing gum, cocktail mixes, herbs, ice cream, plant protein products, seasonings, soft drinks, syrups, vitamin-mineral dietary supplements, and whipped products (imitation).

Use: As a flavor enhancer and flavoring agent.

Health Effects: A large amount of the pure chemical is mildly toxic by swallowing. Eating excessive amounts caused water retention and high blood pressure and loss of potassium, which in turn leads to irregular heartbeat and possible paralysis. GRAS (generally regarded as safe) when used within FDA limitations.

Synonyms: CAS: 8008-94-4 ✲ GLYCYRRHIZA ✲ GLYCYRRHIZAE (LATIN) ✲ GLYCYRRHIZA EXTRACT ✲ GLYCYRRHIZINA ✲ KANZO (JAPANESE) ✲ LICORICE ✲ LICORICE EXTRACT ✲ LICORICE ROOT

LIME, CHLORINATED

Products: In textiles and fabric applications.

Use: For bleaching, deodorizing, and disinfecting.

Health Effects: Product label directions must be followed very carefully.

Synonyms: CHLORIDE OF LIME ✲ BLEACHING POWDER

LIME OIL, DISTILLED

Products: For extracts, flavoring, perfumery, toilet soaps, and cosmetics.

Use: As a flavoring, seasoning, and odorant.

Health Effects: Harmless when used for intended purposes.

Synonyms: CAS: 8008-26-2 ✲ CITRUS AURANTIFOLIA SWINGLE ✲ TERPINEOL ✲ CITRAL

LIME, SULFURATED

Products: In medications, depilatories, luminous paints, and inks.

Use: Various.

Health Effects: An irritant and a caustic.

Synonyms: CALCIUM SULFIDE, CRUDE

LINALOOL

Products: In toilet waters, shaving lotions, hand lotions, and colognes.

Use: As a fruity/floral scented fragrance.

Health Effects: Could cause allergic reactions.

Synonyms: CAS: 78-70-6 ✲ LINALOL ✲ EX BOIS DE ROSE OIL, SYNTHETIC ✲ BERGAMOL OIL ✲ FRENCH LAVENDER

LINOLEIC ACID

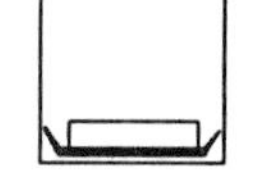

Products: Infant formula and margarine.

Use: As a dietary supplement or flavoring agent.

Health Effects: The pure chemical is a human skin irritant in concentrated form. Swallowing excessive amounts can cause nausea and vomiting. GRAS (generally regarded as safe).

Synonyms: CAS: 60-33-3 ✲ LEINOLEIC ACID ✲ 9,12-LINOLEIC ACID ✲ cis,cis-9,12-OCTADECADIENOIC ACID ✲ cis-9,cis-12-OCTADECADIENOIC ACID ✲ 9,12-OCTADECADIENOIC ACID

LINOLEYTRIMETHYLAMMONIUM BROMIDE

Products: In germicides, deodorants, algicides, and slime control products.

Use: Various.

Health Effects: A severe skin and eye irritant. Highly toxic by swallowing. Label directions should be carefully followed.

Synonyms: None found.

LINSEED OIL

Products: In paints, adhesives, varnishes, putty, printing inks, soaps, and pharmaceutical items.

Use: As a drying agent, thickener, and suspension for various products.

Health Effects: A common allergen and skin irritant.

Synonyms: CAS: 8001-26-1 ❊ GROCO ❊ L-310 ❊ FLAXSEED OIL

LIPOXIDASE

Products: In rolls, buns, breads, and bakery goods.

Use: As a bleaching or whitening agent in bread products.

Health Effects: Harmless when used for intended purposes.

Synonyms: None found.

LITHIUM BROMIDE

Products: In pharmaceuticals, medical sedatives, hypnotic, humectants (attract moisture), and in air conditioning systems.

Use: Various.

Health Effects: Large doses may cause central nervous system depression in humans. Chronic absorption may cause skin eruptions and central nervous system disturbances.

Synonyms: CAS: 7550-35-8

LITHIUM STEARATE

Products: In cosmetics, skin creams, facial makeups, plastics, waxes, and greases. In varnishes, lacquers, and lubricants.

Use: An emulsifier (aids in suspension of oily liquids), lubricant, and plasticizer.

Health Effects: Harmless when used for intended purposes.

Synonyms: None found.

LOCUST BEAN GUM

Products: In baked goods, beverage bases (nonalcoholic), beverages, candy, cheese, fillings, gelatins, ice cream, jams, jellies, pies, puddings, and soups. In cosmetics, textile sizing, textile finishing, and pharmaceuticals.

Use: As an emulsifier (aids in suspension of oily liquids), stabilizer (keeps consistency uniform), and thickening agent.

Health Effects: Mildly toxic by swallowing excessive amounts of concentrated substances. GRAS (generally regarded as safe) within stated FDA limits.

Synonyms: CAS: 9000-40-2 ❊ ALGAROBA ❊ CAROB BEAN GUM ❊ CAROB FLOUR ❊ NCI-C50419 ❊ ST. JOHN'S BREAD ❊ SUPERCOL

LYCOPENE

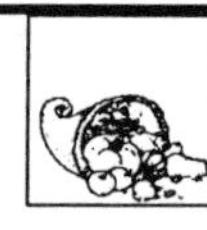

Products: The main pigment of tomato, paprika, and rose hips.

Use: A coloring agent.

Health Effects: Harmless when used for intended purposes.

Synonyms: CAS: 502-65-8 ✲ CAROTENE

D-LYSERGIC ACID DIETHYLAMIDE

Products: Infinitesimal amounts on paper has been put on tongue and can cause deadly detrimental effects.

Use: An illegal, habit-forming hallucinogen. Use often causes unintentional suicide of user.

Health Effects: An illegal controlled substance that is habit forming. It is a strong hallucinogen which elicits optical (visual) or auditory (hearing) hallucinations, depersonalization (person feels mind and body are separated), perceptual disturbances (cannot judge distances), and disturbances of thought processes (cannot control body at will).

Synonyms: CAS: 50-37-3 ✲ LSD

M

"Vitamin E research indicates it is helpful in warding off heart disease, cataracts, strokes, and cancer as well as boosting the immune system."
St. Louis University School of Medicine/Wall Street Journal

MACASSER OIL

Products: In hair tonic, oils, and conditioning treatments.

Use: Oil in grooming products.

Health Effects: Derived from Indian nut kernels. Harmless when used for intended purposes. Could cause allergic reaction in susceptible individuals.

Synonyms: KUSUM OIL ✲ KON OIL ✲ PAKA OIL ✲ CEYLON OAK OIL

MAGNESITE

Products: A white, colorless or grey mineral. In mixes (dry) and table salt.

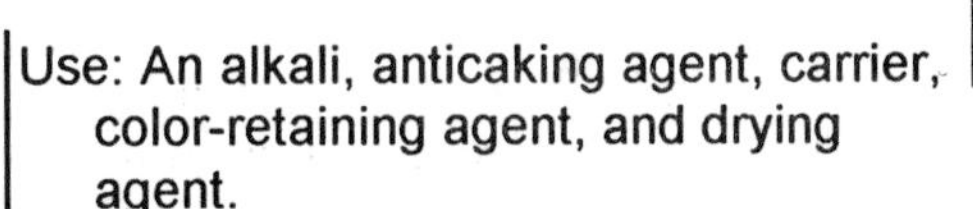

Use: An alkali, anticaking agent, carrier, color-retaining agent, and drying agent.

Health Effects: Harmless when used for intended purposes.

Synonyms: CAS: 546-93-0 ✲ CARBONATE MAGNESIUM ✲ CARBONIC ACID, MAGNESIUM SALT ✲ DCI LIGHT MAGNESIUM CARBONATE ✲ HYDROMAGNESITE ✲ MAGMASTER ✲ MAGNESIA ALBA ✲ MAGNESIUM CARBONATE ✲ MAGNESIUM(II) CARBONATE ✲ MAGNESIUM CARBONATE, PRECIPITATED ✲ STAN-MAG MAGNESIUM CARBONATE

MAGNESIUM

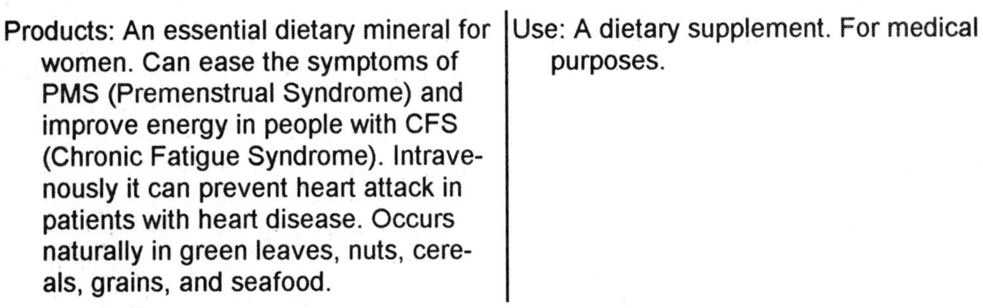

Products: An essential dietary mineral for women. Can ease the symptoms of PMS (Premenstrual Syndrome) and improve energy in people with CFS (Chronic Fatigue Syndrome). Intravenously it can prevent heart attack in patients with heart disease. Occurs naturally in green leaves, nuts, cereals, grains, and seafood.

Use: A dietary supplement. For medical purposes.

Health Effects: Harmless when used for intended purposes in appropriate amounts.

Synonyms: CAS: 7439-95-4 ✲ Mg ✲ ATOMIC NUMBER 12

MAGNESIUM ACETATE

Products: In dyes for textiles, printing, deodorant, disinfectant, and antiseptic.

Use: Various.

Health Effects: Harmless when used for intended purposes.

Synonyms: CAS: 142-72-3 ✲ ACETIC ACID, MAGNESIUM SALT ✲ MAGNESIUM DIACETATE

MAGNESIUM AMMONIUM PHOSPHATE

Products: A fire retardant for fabrics, canvas products, tents, camping equipment, and firefighter apparel.

Use: A fire-proofing agent.

Health Effects: Harmless when used for intended purposes.

Synonyms: MAGNESIUM AMMONIUM ORTHOPHOSPHATE

MAGNESIUM BORATE

Products: An antibacterial and fungus killer.

Use: As a preservative, antiseptic, and fungicide.

Health Effects: Harmless when used for intended purposes.

Synonyms: CAS: 13703-82-7 ✲ ORTHOBORATE ✲ METABORATE

MAGNESIUM CARBONATE

Products: In inks, dentifrices, cosmetics, free-running table salt, antacid, and foods.

Use: As a drying, color-retaining, and anticaking agent. The synthetic form of mineral magnesite.

Health Effects: Harmless when used for intended purposes.

Synonyms: MAGNESITE (NATURALLY OCCURRING MATERIAL)

MAGNESIUM CHLORIDE

Products: For disinfectants, fire extinguishers, fireproofing wood, and floor sweeping compounds. For meat (raw cuts), and poultry (raw cuts).

Use: A color-retention agent; firming agent, flavoring agent, and tissue softening agent (tenderizer) in foods .

Health Effects: Moderately toxic by swallowing the pure chemical. A human mutagen (changes inherited characteristics). In humid environments it causes steel to rust very rapidly. GRAS (generally regarded as safe) when used within the FDA limitations.

Synonyms: CAS: 7786-30-3 ✲ DUS-TOP

MAGNESIUM FLUOSILICATE

Products: Textiles, canvasses, denims, awnings, and fabrics.

Use: For waterproofing, mothproofing, and in laundry products.

Health Effects: Poison by swallowing.

Synonyms: CAS: 18972-56-0 ✲ MAGNESIUM SILICOFLUORIDE ✲ MAGNESIUM HEXAFLUOROSILICATE

MAGNESIUM HYDROXIDE

Products: In antacids, dentifrices, cathartics, and laxatives.

Use: An alkali, color-retaining, drying agent, and for pharmaceutical purposes.

Health Effects: GRAS (generally regarded as safe) when used for intended purposes.

Synonyms: CAS: 1309-42-8 ✲ MAGNESIA MAGMA ✲ MAGNESIUM HYDRATE ✲ MILK of MAGNESIA

MAGNESIUM LAURYL SULFATE

Products: In detergents, cleaning products, and cosmetics.

Use: As a foaming, wetting, and emulsifying (aids in suspension of oily liquids) agent.

Health Effects: Harmless when used for intended purposes.

Synonyms: None found.

MAGNESIUM PERBORATE

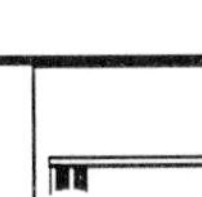

Products: In tooth powders.

Use: For bleaching and antiseptic purposes.

Health Effects: Harmless when used for intended purposes.

Synonyms: CAS: 17097-11-9

MAGNESIUM PEROXIDE

Products: In antacids and medications.

Use: As a bleaching and oxidizing agent.

Health Effects: Harmless when used for intended purposes.

Synonyms: CAS: 1335-26-8 ✲ MAGNESIUM DIOXIDE

MAGNESIUM PHOSPHATE, MONOBASIC

Products: In wood treatment products.

Use: For fireproofing.

Health Effects: Harmless when used for intended purposes.

Synonyms: CAS: 13092-66-5 ✲ MAGNESIUM BIOPHOSPHATE ✲ ACID MAGNESIUM PHOSPHATE ✲ MAGNESIUM TETRAHYDROGEN PHOSPHATE

MAGNESIUM PHOSPHATE, TRIBASIC

Products: An antacid, food additive and dietary supplement; a dentifrice and cleaning agent.

Use: An acid neutralizer, nutritional supplement, and tooth polishing paste.

Health Effects: Harmless when used for intended purposes.

Synonyms: CAS: 7757-86-0 ✲ MAGNESIUM PHOSPHATE, NEUTRAL ✲ TRIMAGNESIUM PHOSPHATE

MAGNESIUM RICINOLEATE

Products: In cosmetic products.

Use: Various.

Health Effects: Could cause allergic reaction in susceptible individuals. Harmless when used for intended purposes.

Synonyms: None found.

MAGNESIUM SILICATE HYDRATE

Products: In table salt.

Use: An anticaking agent and filter aid.

Health Effects: A human skin irritant. GRAS (generally regarded as safe) when used within FDA limitations.

Synonyms: CAS: 1343-90-4

MAGNESIUM STEARATE

Products: In candy, gum, mint, and sugarless gum. Also in dusting powder, medicines, cosmetic emulsifier (maintains mixes and aids in suspension of oils).

Use: An anticaking agent, binder, emulsifier (aids in suspension of oily liquids), lubricant, nutrient, processing aid, release agent, and stabilizer.

Health Effects: Harmless when used for intended purposes. GRAS (generally regarded as safe). Must conform to FDA specifications and within limited amounts.

Synonyms: CAS: 557-04-0

MAGNESIUM SULFATE

Products: For fireproofing textiles; dyeing and printing textiles; mineral waters, dietary supplement, and cosmetic lotions.

Use: Various.

Health Effects: Harmless when used for intended purposes.

Synonyms: CAS: 7587-88-9 ❋ EPSOM SALTS

MALATHION

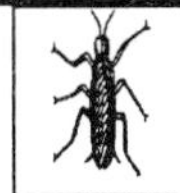

Products: Pesticide in animal feed, cattle feed concentrate blocks (nonmedicated) citrus pulp (dehydrated), grapes, packaging materials, and safflower oil.

Use: As insecticide for flies, insects, head lice, and mosquitos.

Health Effects: A human poison by swallowing and skin contact. Can penetrate intact skin. Effects on the human by swallowing are coma, blood pressure depression, and difficulty in breathing. A possible carcinogen. A human mutagen (changes inherited characteristics). Has caused allergic sensitization of the skin. FDA approves use within limitations.

Synonyms: CAS: 121-75-5 ❋ CALMATHION ❋ CARBETHOXY MALATHION ❋ CARBETOVUR ❋ CARBETOX ❋ CARBOFOS ❋ CARBOPHOS ❋ CELTHIGN ❋ CHEMATHION ❋ CIMEXAN ❋ CYTHION ❋ EMMATOS ❋ EMMATOS EXTRA ❋ S-ESTER with O,O-DIMETHYL PHOSPHOROTHIOATE ❋ ETHIOLACAR ❋ ETIOL ❋ EXTERMATHION ❋ FORMAL ❋ FORTHION ❋ FOSFOTHION ❋ FOSFOTION ❋ FYFANON ❋ HILTHION ❋ KARBOFOS ❋ KOP-THION ❋ KYPFOS ❋ MALACIDE ❋ MALAFOR ❋ MALAGRAN ❋ MALAKILL ❋ MALAMAR ❋ MALAPHELE ❋ MALAPHOS ❋ MALASOL ❋ MALASPRAY ❋ MALATOX ❋ MALDISON ❋ MALMED ❋ MALPHOS ❋ MALTOX ❋ MALTOX MLT ❋ MERCAPTOSUCCINIC ACID DIETHYL ESTER ❋ MERCAPTOTHION ❋ MLT ❋ MOSCARDA ❋ OLEOPHOSPHOTHION ❋ ORTHO MALATHION ❋ PHOSPHORODITHIOIC ACID-O,O-DIMETHYL ESTER-S-ESTER with DIETHYL MERCAPTOSUCCINATE ❋ PHOSPHOTHION ❋ PRIODERM ❋ SADOFOS ❋ SADOPHOS ❋ SUMITOX ❋ TAK ❋ VEGFRU MALATOX ❋ VETIOL ❋ ZITHIOL

MALIC ACID

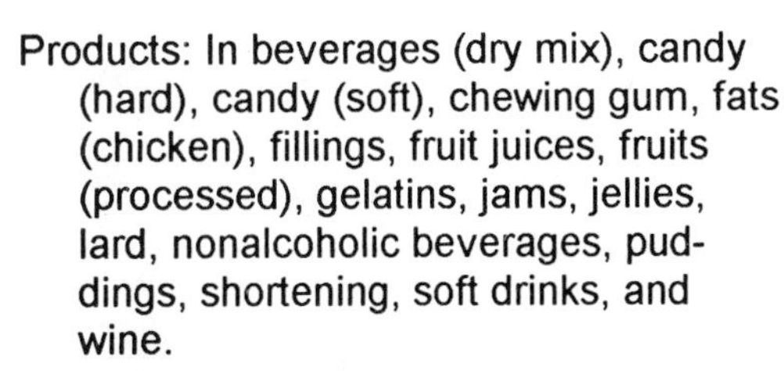

Products: In beverages (dry mix), candy (hard), candy (soft), chewing gum, fats (chicken), fillings, fruit juices, fruits (processed), gelatins, jams, jellies, lard, nonalcoholic beverages, puddings, shortening, soft drinks, and wine.

Use: An acidifier, flavor enhancer, and flavoring agent.

Health Effects: Moderately toxic by swallowing large amounts of the concentrated chemical. A skin and severe eye irritant. GRAS (generally regarded as safe) when used within FDA limitations. Could cause allergic reaction.

Synonyms: CAS: 6915-15-7 ✲ HYDROXYSUCCINIC ACID ✲ KYSELINA JABLECNA

MALTODEXTRIN

Products: In candy, crackers, and puddings.

Use: As a flavor enhancer, bulking agent, crystallization inhibitor, and texturizing agent.

Health Effects: Harmless when used for intended purposes. GRAS (generally regarded as safe).

Synonyms: CAS: 9050-36-6

MALTOL

Products: In baked goods, flour, and wine.

Use: As a flavoring agent and processing aid.

Health Effects: In excessive amounts the pure chemical is moderately toxic by swallowing. A skin irritant. A human mutagen (changes inherited characteristics). FDA approves use when used within stated limitations.

Synonyms: CAS: 118-71-8 ✲ CORPS PRALINE ✲ 3-HYDROXY-2-METHYL-4H-PYRAN-4-ONE ✲ 3-HYDROXY-2-METHYL-γ-PYRONE ✲ 3-HYDROXY-2-METHYL-4-PYRONE ✲ LARIXIC ACID ✲ LARIXINIC ACID ✲ 2-METHYL-3-HYDROXY-4-PYRONE ✲ 2-METHYL-3-OXY-γ-PYRONE ✲ 2-METHYL PYROMECONIC ACID ✲ PALATONE ✲ TALMON ✲ VETOL

MALTOSE

Products: A sweetener formed by action of yeast on starch. Found in bread, instant foods, and pancake syrups.

Use: As a nutritional additive.

Health Effects: Harmless when used for intended purposes.

Synonyms: CAS: 69-79-4 ✲ MALT SUGAR ✲ MALTOBIOSE

MANGANESE GLUCONATE

Products: In baked goods, beverages (nonalcoholic), dairy product analogs, fish products, meat products, milk products, vitamin tablets, and poultry products.

Use: As food additive, dietary supplement, and nutrient.

Health Effects: GRAS (generally regarded as safe).

Synonyms: CAS: 6485-39-8

MANGANESE OCTOATE

Products: In paints, enamels, varnishes, and inks.

Use: As a drying agent.

Health Effects: Harmless when used for intended purposes.

Synonyms: MANGANESE ETHYLEXOATE

MANNITOL

Products: For sweetener; dietetic food base,

Use: As a thickener, stabilizer, flavoring agent, and lubricant.

Health Effects: Mildly toxic by swallowing excessive amounts of the pure substance. A human mutagen (changes inherited characteristics).

Synonyms: CAS: 69-65-8 ❊ MANNA SUGAR ❊ MANNITE ❊ OSMITROL ❊ 1,2,3,4,5,6-HEXANEHEXOL

MARIJUANA

Products: Smoking materials.

Use: Hallucinogenic illegal street drug. Does have medical applications for controlling nausea.

Health Effects: An hallucinogen. Moderately toxic by swallowing. An animal teratogen (causes abnormal fetus development), causes reproductive effects (infertility, or sterility, or birth defects). Human mutagen (changes inherited characteristics). Harmful effects include heart rate changes, blood pressure drop. When smoked or swallowed can cause delirium, drowsiness, weakness, reflex weakness. Overdose can cause coma and death. This herb does have medical applications and can be very beneficial for epileptics and controlling nausea of chemotherapy patients.

Synonyms: CAS: 8063-14-7 ❊ CANNABIS ❊ MARY JANE ❊ DOPE ❊ INDIAN HEMP ❊ HASHISH

MARJORAM OIL

Products: In perfume and toilet waters. The Spanish grade is used in fish, meat, sauces, and soups.

Use: As a flavoring agent and odorant.

Health Effects: GRAS (generally regarded as safe) when used at moderate levels to accomplish the intended effect.

Synonyms: None found.

MENTHOL

Products: In cigarettes, perfumes, shaving creams, lotions, after shave, hair products, foods, liqueurs, chewing gum, cold medications, and cough drops.

Use: As a flavoring agent and odorant.

Health Effects: In large amounts it is moderately toxic by swallowing. A severe eye irritant. FDA approves use at moderate levels to accomplish the intended effect.

Synonyms: CAS: 89-78-1 ✲ HEXAHYDROTHYMOL ✲ 2-ISOPROPYL-5-METHYL-CYCLOHEXANOL ✲ p-MENTHAN-3-OL ✲ 1-MENTHOL ✲ 5-METHYL-2-(1-METHYLETHYL)CYCLOHEXANOL ✲ PEPPERMINT CAMPHOR

MENTHONE

Products: In beverages, ice creams, candies, bakery products, and chewing gum.

Use: As a fruit or mint flavoring.

Health Effects: Moderately toxic by swallowing excessive amounts of chemical. A mutagen (changes inherited characteristics). FDA approves use at moderate levels to accomplish the intended effect.

Synonyms: CAS: 89-80-5 ✲ l-p-MENTHAN-3-ONE ✲ l-MENTHONE (FCC) ✲ p-MENTHONE ✲ trans-MENTHONE ✲ trans-5-METHYL-2-(1-METHYLETHYL)-CYCLOHEXANONE

MERCERIZED COTTON

Products: For threads, yarns, and cottons.

Use: A strengthening process by which material is passed through solution of sodium hydroxide and washed with water while under tension. Causes fibers to shrink, resulting in strengthening and improving color properties.

Health Effects: Harmless when used for intended purposes.

Synonyms: None found.

MERCURIC ARSENATE

Products: In paints.

Use: For waterproofing and marine antifouling (prevents underwater growth) paint.

Health Effects: A poison. A confirmed carcinogen (causes cancer). Label instructions for use must be strictly followed.

Synonyms: CAS: 7784-37-4 ❖ MERCURY ARSENATE ❖ MERCURY ORTHOARSENATE

MERCURIC CHLORIDE

Products: In fungicide, insecticide, wood preservative, embalming fluid, and printing.

Use: Kills fungus and bacteria.

Health Effects: Toxic by swallowing, breathing, and skin absorption. A poison.

Synonyms: CAS: 7487-94-7

MERCURIC CYANIDE

Products: In antiseptics and germicidal soaps.

Use: For cleansing and antibacterial products.

Health Effects: Toxic by swallowing, breathing, and skin absorption. Effects on the human body by swallowing are nausea, vomiting, diarrhea, and somnolence (sleepiness).

Synonyms: CAS: 592-04-1 ❖ MERCURY CYANIDE

MERCURIC NAPHTHENATE

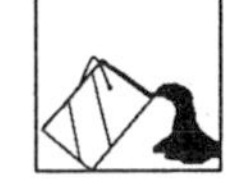

Products: In paints, stains, and similar products.

Use: A mold and mildew-resistance agent.

Health Effects: Toxic by swallowing, breathing, and skin absorption.

Synonyms: MERCURY NAPHTHENATE

MERCURIC OLEATE

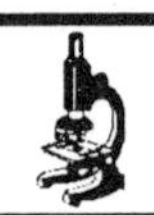

Products: In antiseptics; antifouling paints.

Use: To kill mold, mildew, and bacteria.

Health Effects: A poison. An FDA over-the-counter drug.

Synonyms: CAS: 1191-80-6 ❖ MERCURY OLEATE

MERCURIC OXIDE, RED

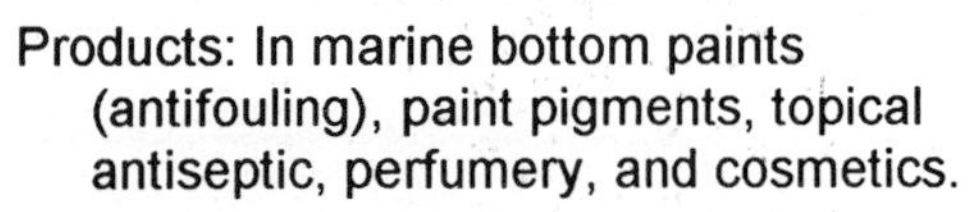

Products: In marine bottom paints (antifouling), paint pigments, topical antiseptic, perfumery, and cosmetics.

Use: As a coloring, polishing, and anti-bacterial compound.

Health Effects: On EPA extremely hazardous substance list. Poison by swallowing and skin contact. An FDA over-the-counter drug.

Synonyms: CAS: 21908-53-2 ❖ RED PRECIPITATE ❖ MERCURY OXIDE, RED ❖ SANTAR ❖ YELLOW PRECIPITATE ❖ RED OXIDE OF MERCURY

MERCURIC SODIUM, PHENOSULFONATE

Products: In soaps, lotions, and ointments.

Use: As antiseptic and germicide.

Health Effects: Toxic by swallowing and skin absorption.

Synonyms: CAS: 535-55-7 ✲ HERMOPHENYL ✲ MERCURY AND SODIUM PHENOSULFONATE

MESCALINE

Products: Derived from a Mexican cactus.

Use: In medical and biochemical research.

Health Effects: An hallucinogenic drug. Moderately toxic by swallowing. Poison by intravenous route. Effects on the body by intramuscular route are hallucinations and distorted perceptions,

Synonyms: CAS: 54-04-6 ✲ MEZCALINE ✲ MEZCLINE ✲ 3,4,5-TRIMETHOXYBENZENEETHANIMINE ✲ 3,4,5-TRIMETHOXYPHENETHYLAMINE

MESITYL OXIDE

Products: In lacquers, inks, stains, paint, varnish remover, and insecticide.

Use: As solvent, remover, and repellent.

Health Effects: Moderately toxic by swallowing. Mildly toxic by breathing and skin contact. Irritating to the eyes. High concentrations are narcotic. Readily absorbed through skin. Prolonged exposure can injure liver, kidney, and lungs.

Synonyms: CAS: 141-79-7 ✲ METHYL ISOBUTENYL KETONE ✲ ISOPROPYLIDENEACETONE ✲ 4-METHYL-3-PENTEN-2-ONE

METEPA

Products: An insect chemosterilant (prevents reproduction). On permanent press fabrics to prevent wrinkling and for fireproofing. In adhesives.

Use: Various.

Health Effects: Toxic by swallowing and skin absorption. A strong skin irritant.

Synonyms: CAS: 57-39-6 ✲ TRIS(METHYL-1-AZIRIDINYL)PHOSPHINE OXIDE ✲ METHYL APHOXIDE ✲ MAPO ✲ METHAPHOXIDE

METHANE

Products: Natural gas and coal gas from decaying vegetation and other organic

Use: In the form of natural gas, methane is used as a fuel. It is purchased by matter in swamps and marshes. power companies from dump sites for home heating and cooking.

Health Effects: Severe fire and explosion hazard; forms explosive mixture with air (5-15% by volume).

Synonyms: CAS: 74-82-8 ✲ MARSH GAS ✲ METHYL HYDRIDE

METHIONINE

Products: An amino acid essential in human nutrition. A dietary supplement and nutrient. A cosmetic cream and lotion texturizer.

Use: As vegetable oil enricher and toiletry conditioner.

Health Effects: Mildly toxic by swallowing excessive quantities. A possible human mutagen (changes inherited characteristics).

Synonyms: CAS: 63-68-3 ✲ CYMETHION ✲ LIQUIMETH

METHOXSALEN

Products: In skin products for suntan lotions, gels, mousses, creams, and ointments.

Use: In suntan accelerator and sunburn protector.

Health Effects: A confirmed carcinogen (causes cancer). Moderately toxic by swallowing. A human mutagen (changes inherited characteristics).

Synonyms: CAS: 298-81-7 ✲ XANTHOTOXIN ✲ AMMOIDIN ✲ MELADININ ✲ MELOXINE ✲ METHOXA-DOME ✲ 8-METHOXYPSORALEN ✲ OXSORALEN ✲ OXYPSORALEN ✲ PRORALONE-MOP

METHYL ABIETATE

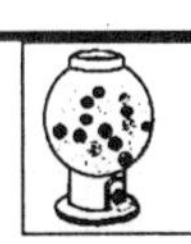

Products: Masticatory substance in chewing gum base. In lacquers, varnishes, adhesives, and coatings.

Use: As a solvent, softener, and plasticizer.

Health Effects: A skin irritant. Probably slightly toxic.

Synonyms: CAS: 127-25-3 ✲ ABIETIC ACID, METHYL ESTER ✲ METHYL ESTER of WOOD ROSIN ✲ METHYL ESTER of WOOD ROSIN, partially hydrogenated (FCC)

METHYL ACETATE

Products: In paint remover compounds, lacquer solvent. In manufacture of artificial leathers. A perfume enhancer.

Use: Various.

Health Effects: Flammable, dangerous fire and explosion risk. Irritant to respiratory tract. Narcotic in high concentrations. A moderate skin and severe eye irritant.

Synonyms: CAS: 79-20-9 ✲ ACETIC ACID METHYL ESTER ✲ DEVOTON ✲ METHYL ETHANOATE ✲ TERETON

METHYL ACETONE

Products: In lacquers, paint, and varnish removers. Perfume extraction.

Use: As a solvent.

Health Effects: Flammable, dangerous fire risk. Toxic by swallowing.

Synonyms: ACETONE ETHYL ACETATE METHANOL

METHYLAL

Products: In perfumes, adhesives, protective coatings, and fuels.

Use: Various.

Health Effects: Toxic by swallowing and breathing.

Synonyms: CAS: 109-87-5 ✲ FORMAL ✲ DIMETHOXYMETHANE

METHYL ALCOHOL

Products: In antifreeze, solvents, shellacs, dyes, utility plant fuel, and home heating oil.

Use: As deicer, solvent, and oil extender.

Health Effects: A human poison by swallowing. Mildly toxic by breathing. Effects on the human body are changes in circulation, cough, dyspnea (shortness of breath), headache, eyes watering, nausea, vomiting, blindness, and respiratory effects. A narcotic. A human mutagen (changes inherited characteristics). Flammable, dangerous fire risk. A cumulative poison.

Synonyms: CAS: 67-56-1 ✲ METHANOL ✲ WOOD ALCOHOL ✲ CARBINOL ✲ METHYL HYDROXIDE ✲ PYROXYLIC SPIRIT ✲ WOOD NAPHTHA ✲ METHYLOL ✲ WOOD SPIRIT

METHYLAMINE

Products: In dyes, pharmaceuticals, fungicides, fuel additive, insecticide, and fungicide.

Use: In paint remover, solvent, pesticide, and fungus destroyer.

Health Effects: Moderately toxic by breathing. A severe skin irritant. A human mutagen (changes inherited characteristics).

Synonyms: CAS: 74-89-5 ✲ MONOMETHYLAMINE ✲ AMINOMETHANE ✲ CARBINAMINE ✲ MERCURIALIN

METHYL AMYL KETONE

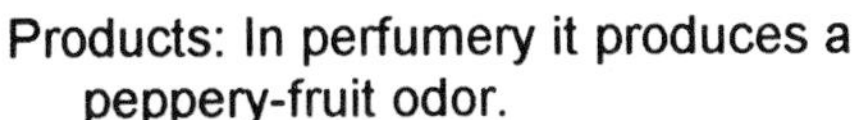

Products: In perfumery it produces a peppery-fruit odor.

Use: In lacquers, solvents, and as a flavoring agent.

Health Effects: Moderately toxic by swallowing. Mildly toxic by breathing and skin contact. A skin irritant.

Synonyms: CAS: 110-43-0 ✽ n-AMYL METHYL KETONE ✽ AMYL METHYL KETONE ✽ 2-HEPTANONE ✽ METHYL-AMYL-CETONE ✽ METHYL AMYL KETONE ✽ METHYL PENTYL KETONE

METHYL ANTHRANILATE

Products: A grape-like, fruity, flavoring. An odorant in skin cosmetics and hair pomades.

Use: A perfume and food flavoring agent. An ingredient that affects the taste or smell of final product.

Health Effects: Moderately toxic by swallowing excessive quantities of pure chemical. A skin irritant. Could cause allergic reactions in susceptible individuals.

Synonyms: CAS: 134-20-3 ✽ METHYL-o-AMINOBENZOATE ✽ NEROLI OIL, ARTIFICIAL

METHYL BROMIDE

Products: A powerful fumigant gas.

Use: To fumigate homes and other buildings under tenting. To destroy infestations of insects in fruits and vegetables. To fumigate import shipments.

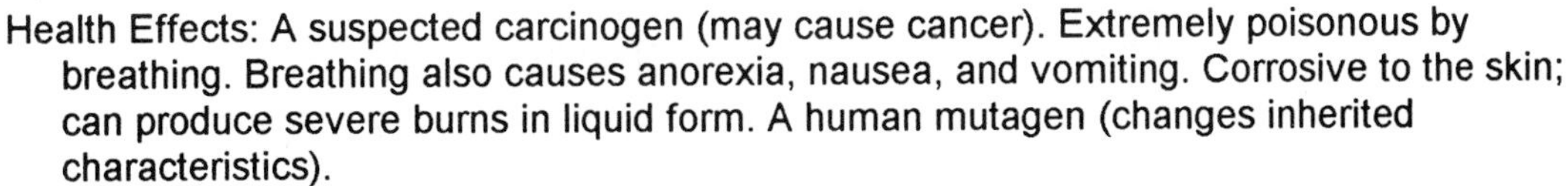

Health Effects: A suspected carcinogen (may cause cancer). Extremely poisonous by breathing. Breathing also causes anorexia, nausea, and vomiting. Corrosive to the skin; can produce severe burns in liquid form. A human mutagen (changes inherited characteristics).

The effects are cumulative and damaging to the nervous system, kidneys, and lungs. Central nervous system effects include blurred vision, mental confusion, numbness, tremors, and speech defects.

Synonyms: CAS: 74-83-9 ✽ BROMOMETHANE ✽ DOWFUME ✽ HALON 1001 ✽ ISCOBROME ✽ METAFUME ✽ METAGAS ✽ METHOGAS ✽ PROFUME

METHYL CAPROATE

Products: In skin creams, detergents, and lubricants.

Use: A stabilizer and emulsifier (maintains mixes to aid in suspension of oily liquids).

Health Effects: Harmless when used for intended purposes.

Synonyms: CAS: 106-70-7 ❖ METHYL HEXANOATE ❖ METHYL ESTER CAPROIC ACID

METHYLBENZETHONIUM CHLORIDE

Products: A bactericide for external skin lesions and abrasions.

Use: A germ killing medication.

Health Effects: Mildly toxic by swallowing.

Synonyms: CAS: 25155-18-4 ❖ BACTINE

METHYL CELLULOSE

Products: In adhesives and paint pigments. In baked goods, fruit pie fillings, meat patties, vegetable patties, and diet foods.

Use: As a binder, bodying agent, bulking agent, emulsifier (aids in suspension of oily liquids), film former, stabilizer, thickening, and sizing agent.

Health Effects: GRAS (generally regarded as safe) when used within FDA limitations.

Synonyms: CAS: 9004-67-5 ❖ CELLULOSE METHYL ETHER ❖ METHOCEL

METHYL CHLORIDE

Products: A refrigerant, herbicide, insecticideal propellant, local, or topical anesthetic (acts by freezing tissue).

Use: As solvent, weed killer, insect killer; has skin numbing effects.

Health Effects: Suspected carcinogen (causes cancer). Very mildly toxic by breathing. A human mutagen (changes inherited characteristics). Effects on the human body by breathing are: convulsions, nausea or vomiting, and unspecified effects on the eye.

Synonyms: CAS: 74-87-3 ❖ ARTIC ❖ CHLOROMETHANE ❖ MONOCHLOROMETHANE

METHYLENE CHLORIDE

Products: In coffee (decaffeinated), fruits, hops extract, spice oleoresins, vegetables, adhesives, glues, cleaners, waxes, oven cleaners, paint strippers, paint removers, shoe polish, varnish, stain, and sealant.

Use: For a degreasing and cleaning fluid; as a solvent for food processing. For color dye or fixative; as an extraction solvent. Not used much as a propellant anymore.

Health Effects: A confirmed carcinogen (causes cancer). Moderately toxic by swallowing concentrated amounts. Mildly toxic by breathing. Effects on the human body by swallowing concentrated amounts and breathing are paresthesia (abnormal sensation of burning or tingling), somnolence (sleepiness), altered sleep time, convulsions, euphoria, and change in cardiac rate. An eye and severe skin irritant. A human mutagen (changes inherited characteristics). FDA approves use within limitations.

Synonyms: CAS: 75-09-2 ❖ AEROTHENE MM ❖ DCM ❖ DICHLOROMETHANE ❖ FREON 30 ❖ METHANE DICHLORIDE ❖ METHYLENE BICHLORIDE ❖ METHYLENE DICHLORIDE ❖ SOLMETHINE

METHYL ETHYL CELLULOSE

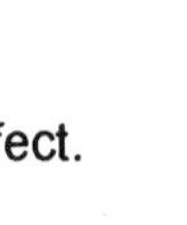

Products: In meringues and whipped toppings. Pharmaceutically in laxative products

Use: An emulsifier, foaming agent, stabilizer, aerator, and bowel bulking agent.

Health Effects: FDA approves use at moderate levels to accomplish the intended effect.

Synonyms: CELLULOSE, MODIFIED

METHYL ETHYL KETONE

Products: In paint removers, cements, and cleaning fluids.

Use: As solvent, adhesive, surface-coating, and a stain remover.

Health Effects: Moderately toxic by swallowing and skin contact. Effects on the human body by breathing: nose, throat, conjunctiva (around eye) irritation, and unspecified effects on the respiratory system. A strong irritant. Affects peripheral nervous system and central nervous system.

Synonyms: CAS: 78-93-3 ❖ 2-BUTANONE ❖ ETHYL METHYL KETONE ❖ MEK ❖ METHYL ACETONE

N-METHYLGLUCAMINE

Products: In detergents, pharmaceuticals, and dyes.

Use: Various.

Health Effects: Can cause allergic reactions to susceptible individuals.

Synonyms: CAS: 6284-40-8 ❖ GLUCOSE METHYLAMINE

METHYL GLYCOL

Products: In perfumes, colors, soft drink syrups, flavoring extracts, cleansing creams, fabric softeners, sun tan lotions, brake fluids, antifreeze, coolants, deicers, and tobacco.

Use: As a solvent, conditioner, wetting agent, humectant (keeps product from drying out), emulsifier (stabilizes and maintains mixes), anticaking agent, preservative, mold and fungus retarder. As a spray mist to disinfect the air.

Health Effects: Slightly toxic by swallowing concentrated amounts and excessive skin contact. Effects on the human body by swallowing large amounts are convulsions and general anesthesia. An eye and skin irritant.

Synonyms: CAS: 57-55-6 ❖ 1,2-PROPYLENE GLYCOL ❖ 1,2-DIHYDROXYPROPANE ❖ 1,2-PROPANEDIOL ❖ METHYLENE GLYCOL ❖ POLYPROPYLENE GLYCOL

METHYLHEPTENONE

Products: Inexpensive perfume and odorants. In flavorings.

Use: For citrus-lemon fragrance and taste.

Health Effects: Moderately toxic by swallowing excessive amounts. A skin irritant.

Synonyms: CAS: 409-02-9 ❖ 6-METHYL-5-HEPTENE-2-ONE

METHYL HEXYL KETONE

Products: In perfumes, solvents, coatings, and leather finishes.

Use: As a flavoring and odorant.

Health Effects: A skin irritant.

Synonyms: CAS: 111-13-7 ❖ 2-OCTANONE

METHYL-12-HYDROXYSTEARATE

Products: In adhesives, inks, cosmetics, ointments, and greases.

Use: A waxy thickening and suspension material.

Health Effects: Could cause allergic reaction in susceptible individuals.

Synonyms: CAS: 141-23-1 ❖ 12 HYDROXYSTEARIC ACID, METHYL ESTER

N-METHYL MORPHOLINE

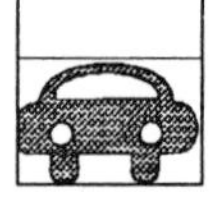

Products: In self-polishing waxes and pharmaceuticals.

Use: Various.

Health Effects: Moderately toxic by swallowing and skin contact. Mildly toxic by breathing. A corrosive irritant to skin, eyes, nose, and throats.

Synonyms: CAS: 109-02-4 ❖ METHYLMORPHOLINE ❖ 4-METHYLMORPHOLINE

METHYLOACRYLAMIDE

Products: In varnishes, adhesives, crease-proof fabrics, wrinkle-resistant fabrics, and permanent press textiles.

Use: For coating, cementing, and textile treatment.

Health Effects: Harmless when used for intended purposes. Some individuals experience allergic skin reactions when they come in contact with these materials.

Synonyms: None found.

METHYLPARABEN

Products: A preservative in cosmetics, facial moisturizers, lipstick, eye shadow, mascara, and others.

Use: An antimicrobial (germ killing) agent to prevent bacterial growth in makeup products.

Health Effects: Could cause allergic reaction. Moderately toxic by swallowing. A mutagen (changes inherited characteristics).

Synonyms: CAS: 99-76-3 ✲ PARABENS ✲ METHYL-p-HYDROXYBENZOATE ✲ ABIOL ✲ ASEPTOFORM ✲ METHYLBEN ✲ METHYL CHEMOSEPT ✲ METHYL PARASEPT ✲ METOXYDE ✲ MOLDEX ✲ PARASEPT ✲ PARIDOL ✲ SEPTOS ✲ PRESERVAL M ✲ TEGOSEPT M

METHYL PHENYLACETATE

Products: Honey-like odor for perfumery, flavors for tobacco, and flavoring.

Use: Ingredient to affect taste and/or smell of products.

Health Effects: Could cause allergic reaction in susceptible individuals.

Synonyms: None found.

METHYLPHOSPHORIC ACID

Products: In rust remover.

Use: For steel finishes on vehicles, farm equipment, bicycles, toys, etc.

Health Effects: Directions must be closely followed on products that contain this chemical.

Synonyms: METHYL ORTHOPHOSPHORIC ACID ✲ METHYL ACID PHOSPHATE

METHYL SALICYLATE

Products: As a UV (ultaviolet) absorber in sunburn lotions. Flavoring in foods, baked goods, beverages, candy, odorant, perfumery, chewing gum, and topical analgesic (pain killer).

Use: Various.

Health Effects: Human poison by swallowing large quantities of concentrated chemical. Effects in the human body by swallowing are dyspnea (shortness of breath), nausea, vomiting, and respiratory stimulation. A severe skin and eye irritant. Swallowing of relatively small amounts has caused severe poisoning and death.

Synonyms: CAS: 119-36-8 ✲ o-ANISIC ACID ✲ BETULA OIL ✲ GAULTHERIA OIL, ARTIFICIAL ✲ o-HYDROXYBENZOIC ACID, METHYL ESTER ✲ 2-HYDROXYBENZOIC ACID METHYL ESTER ✲ o-METHOXYBENZOIC ACID ✲ 2-METHOXYBENZOIC ACID ✲ METHYL-o-HYDROXYBENZOATE ✲ NATURAL WINTERGREEN OIL ✲ OIL of WINTERGREEN ✲ SALICYLIC ACID, METHYL ESTER ✲ SWEET BIRCH OIL ✲ SYNTHETIC WINTERGREEN OIL ✲ TEABERRY OIL ✲ WINTERGREEN OIL ✲ WINTERGREEN OIL, SYNTHETIC

β-METHYLUMBELLIFERONE

Products: In soaps, starches, laundry products, and suntan lotions.

Use: Produces a bright, blue-white fluorescence in daylight or UV (ultraviolet) light. Causes white fabrics to appear brighter and cleaner.

Health Effects: Could cause allergic reaction on skin contact with treated materials.

Synonyms: BMU ✲ 7-HYDROXY-4-METHYL COUMARIN

MILK OF MAGNESIA

Products: Laxative medication.

Use: As a purgative.

Health Effects: Harmless when used for intended purposes.

Synonyms: MAGNESIA MAGMA ✲ MAGNESIUM HYDROXIDE

MINERAL OIL

Products: In bakery products, beet sugar, confectionery, egg white solids, fruit (raw), fruits (dehydrated), grain, meat (frozen), pickles, potatoes (sliced), sorbic acid, starch (molding), vegetables (dehydrated), vegetables (raw), vinegar, wine, and yeast. Also a laxative. In cosmetic creams, baby lotions, hair conditioners, lipstick, mascara, blush, shaving creams, makeup foundations, cleansing creams, and suntan lotions.

Use: As a binder, defoaming agent, fermentation aid, lubricant, coating (protective), and release agent.

Health Effects: A petroleum derivative. In excessive amounts the concentrated chemical is a human teratogen (abnormal fetal development) by breathing, which causes testicular tumors in the fetus. Breathing of vapor or particulates can cause aspiration pneumonia. A skin and eye irritant. A possible carcinogen (causes cancer) producing gastrointestinal (stomach and intestines) tumors. Combustible liquid when exposed to heat or flame. There is also a purified food grade which is approved for use within limitations by the FDA.

Synonyms: CAS: 8012-95-1 ✲ ADEPSINE OIL ✲ ALBOLINE ✲ BAYOL F ✲ BLANDLUBE ✲ CRYSTOSOL ✲ DRAKEOL ✲ FONOLINE ✲ GLYMOL ✲ KAYDOL ✲ KONDREMUL ✲ MINERAL OIL, WHITE ✲ MOLOL ✲ NEO-CULTOL ✲ NUJOL ✲ OIL MIST, MINERAL ✲ PAROL ✲ PAROLEINE ✲ PARRAFIN OIL ✲ PENETECK ✲

PENRECO ❋ PERFECTA ❋ PETROGALAR ❋ PETROLATUM, liquid ❋ PRIMOL 335 ❋ PROTOPET ❋ SAXOL ❋ TECH PET F ❋ WHITE MINERAL OIL

MONAMINES

Products: In detergents, detergent additives, soaps, shampoos, cleansers, and cleaning agents.

Use: As foam boosters, wetters, (breaks down soil and grease), dispersing agents (mixes and spreads ingredients), thickeners, and conditioners.

Health Effects: Could cause allergic reaction in susceptible individuals.

Synonyms: DIALKYLOLAMIDES

MONOAMMONIUM GLUTAMATE

Products: In meat and poultry.

Use: A flavor enhancer and salt substitute.

Health Effects: FDA states GRAS (generally regarded as safe) when used at moderate amounts to accomplish the intended purpose.

Synonyms: CAS: 7558-63-6 ❋ AMMONIUMGLUTAMINAT ❋ MAG ❋ MONOAMMONIUM l-GLUTAMATE

MONOGLYCERIDE

Products: In cosmetics, creams, lotions, facial rouges, mascaras, and eye shadows.

Use: As an emulsifier (stabilizes and maintains mixes to aid in suspension of oily liquids), and lubricant.

Health Effects: Could cause allergic reaction in susceptible individuals.

Synonyms: GLYCEROL MONOSTEARATE ❋ GLYCEROL MONOLAURATE

MONO- and DIGLYCERIDES

Products: In baked goods (yeast raised), cakes, caramel, chewing gum, coffee whiteners, fats (rendered animal), frozen desserts, fudge, fudge sauces, lard, oleomargarine, peanut butter, poultry, shortening, and whipped topping.

Use: As a dough strengthener, emulsifier (maintains mixes and aids in suspension of ingredients), emulsifier salt, flavoring agent, formulation aid, lubricant, release agent, softener, solvent, stabilizer, texturizing agent, thickening agent, or vehicle.

Health Effects: FDA states GRAS (generally regarded as safe) when used within limitations.

Synonyms: None found.

MONO- and DIGLYCERIDES, MONOSODIUM PHOSPHATE DERIVATIVES

Products: In candy (soft) and simulated dairy products.

Use: As an emulsifier (maintains mixes and suspends ingredients), emulsifier salt, lubricant, and a release agent.

Health Effects: FDA states GRAS (generally regarded as safe).

Synonyms: MONOSODIUM PHOSPHATE DERIVATIVES of MONO- and DIGLYCERIDES

MONO-, DI-, and TRISTEARYL CITRATE

Products: In oleomargarine.

Use: As a flavor preservative, stabilizer, or sequestrant (affects the final products appearance, flavor, or texture).

Health Effects: FDA approves use within limits.

Synonyms: None found.

MONOGLYCERIDE CITRATE

Products: In fats, fats (poultry), lard, margarine, oil, oleomargarine, sausage (fresh pork), meat (dried), and shortening.

Use: As an antioxidant (slows spoiling); synergist (aids in action) for antioxidants.

Health Effects: FDA approves use within stated limitations.

Synonyms: None found.

MONOSODIUM GLUTAMATE

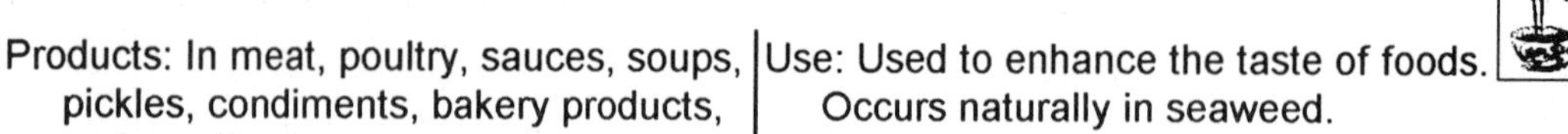

Products: In meat, poultry, sauces, soups, pickles, condiments, bakery products, and candies.

Use: Used to enhance the taste of foods. Occurs naturally in seaweed.

Health Effects: Mildly toxic by swallowing excessive amounts of pure substance. Effects on the body by swallowing are somnolence (sleepiness), hallucinations, distorted perceptions, headache, dyspnea (shortness of breath), nausea, vomiting, and dermatitis. The cause of "Chinese restaurant syndrome". Some people experience mood changes, irritability, numbness, and depression. It can trigger IBS (Irritable Bowel Syndrome). Animal studies resulted in reproductive effects (infertility, or sterility, or birth defects) and teratogenic (abnormal fetal development) effects. FDA permits use within limitations. It has a pharmacologic effect. This means that given enough M. S. G. anyone would develop symptoms. This is different than an allergic effect. It could be compared to the effect of alcohol on individuals. That is, people have different tolerances to it.

Synonyms: CAS: 142-47-2 ✲ ACCENT ✲ AJINOMOTO ✲ CHINESE SEASONING ✲ GLUTACYL ✲ GLUTAMIC ACID, SODIUM SALT ✲ GLUTAVENE ✲ MONOSODIUM-*l*-GLUTAMATE ✲ α-MONOSODIUM GLUTAMATE ✲ MSG ✲ RL-50 ✲ SODIUM GLUTAMATE ✲ SODIUM *l*-GLUTAMATE ✲ *l*-(+) SODIUM GLUTAMATE ✲ VETSIN ✲ ZEST

MONTAN WAX

Products: In shoe polish, furniture polish, roof paints, waterproof paints, adhesives, pastes, carbon paper, and substitute for carnauba and beeswax.

Use: As protective ingredient and decorative coating.

Health Effects: Harmless when used for intended purposes.

Synonyms: LIGNITE WAX

MORPHINE

Products: Analgesics (to reduce or prevent mortal pain).

Use: Medical.

Health Effects: Narcotic; a habit-forming drug. Sale is restricted by law in the U.S. Poison by swallowing, subcutaneous (injected under the skin), and intravenous (injected in the vein). Causes reproductive (infertility, sterility, or birth defects) effects.

Synonyms: CAS: 57-27-2 ✲ MORPHIA ✲ MORPHINA ✲ MORPHINIUM ✲ MORPHIUM

MORPHOLINE

Products: In waxes, polishes, detergent brightener, pharmaceuticals, bactericides, local anesthetics, and antiseptics.

Use: Various.

Health Effects: Moderately toxic by swallowing, breathing, and skin contact. A human mutagen (changes inherited characteristics). A corrosive irritant to skin, eyes, nose, and throat. Can cause kidney damage. A possible carcinogen (causes cancer).

Synonyms: CAS: 110-91-8 ✲ TETRAHYDRO-1,4-OXAZINE ✲ DIETHYLENEIMIDE OXIDE ✲ DIETHYLENE IMIDOXIDE ✲ DIETHYLENE OXIMIDE

MUCILAGE

Products: Glues, adhesives, and gum.

Use: Used for adhesion. Derived from seeds, roots, plants, algae, or seaweed.

Health Effects: Harmless when used for intended purposes.

Synonyms: ALGAL POLYSACCHARIDES ❖ AGAR MUCILAGE ❖ ALGIN MUCILAGE ❖ CARRAGEENIN MUCILAGE

MUSK

Products: In cosmetics, shaving lotions, shaving creams, colognes, toilet waters, air fresheners, and mothproofers.

Use: As a fragrance or odorant.

Health Effects: Human poison by an unspecified route. Can cause allergic reaction in susceptible individuals.

Synonyms: CAS: 300-54-9 ❖ MUSCARINE ❖ MUSCARIN ❖ MUSKARIN ❖ *dl*-MUSCARINE

MYRISTIC ACID

Products: In cosmetic, soaps, and shaving creams

Use: A defoaming agent and lubricant.

Health Effects: A human mutagen (changes inherited characteristics). A human skin irritant. Could cause allergic reactions.

Synonyms: CAS: 544-63-8 ❖ CRODACID ❖ EMERY 655 ❖ HYDROFOL ACID 1495 ❖ HYSTRENE 9014 ❖ TETRADECANOIC ACID ❖ n-TETRADECOIC ACID ❖ 1-TRIDECANECARBOXYLIC ACID ❖ UNIVOL U 316S

MYRISTYL ALCOHOL

Products: In soaps, cosmetics, detergents, ointments, suppositories, shampoos, toothpastes, cold creams, and cleaning preparations.

Use: A fixative, wetting agent, and antifoam agent.

Health Effects: A human skin irritant.

Synonyms: CAS: 112-72-1 ❖ 1-TETRADECANOL ❖ TETRADECYL ALCOHOL

MYRRH OIL

Products: In incense, perfumes, dentifrices, foods, and beverages.

Use: As an ingredient to affect the taste or smell of products.

Health Effects: Harmless when used for appropriate purposes. FDA approves use at moderate levels to accomplish the intended effect.

Synonyms: CAS: 9000-45-7

MYXIN

Products: An antibiotic, bacteriostat, and

Use: As germicide or fungicide.

antifungal.

Health Effects: Harmless when used for intended effect.

Synonyms: CAS: 13925-12-7 ❖ 6-METHOXY-1-PHENAZINOL-5,10-DIOXIDE

N

40% of U. S. winter produce comes from outside the country where U. S. regulations regarding residual pesticide may not be followed.

NALED

Products: Insecticides. Acaricide (kills worms).

Use: As pesticide.

Health Effects: Poison by swallowing and breathing. Moderately toxic by skin contact. A skin irritant.

Synonyms: CAS: 300-76-5 ❖ ARTHODIBROM ❖ BROMCHLOPHOS ❖ BROMEX ❖ DIBROM ❖ 1,2-DIBROMO-2,2-DICHLOROETHYL DIMETHYL PHOSPHATE ❖ O,O-DIMETHYL-O-(1,2-DIBROMO-2,2-DICHLOROETHYL)PHOSPHATE ❖ DIMETHYL-1,2-DIBROMO-2,2-DICHLOROETHYL PHOSPHATE ❖ O,O-DIMETHYL-O-2,2-DICHLORO-1,2-DIBROMOETHYL PHOSPHATE ❖ HIBROM ❖ ORTHO 4355 ❖ ORTHODIBROM ❖ ORTHODIBROMO

NAPHTHA

Products: In paints, stains, finishes, varnish thinner, dry cleaning fluid, asphalt solvent, and naphtha soaps. For eggs (shell), fruits (fresh), and vegetables (fresh).

Use: As a coloring agent, coating (protective), solvent, and various other uses.

Health Effects: Mildly toxic by breathing. Can cause unconsciousness which may go into coma, labored breathing, and bluish tint to the skin. Recovery follows removal from exposure. In mild form, intoxication resembles drunkenness. On a chronic basis, no true poisoning; sometimes headache, lack of appetite, dizziness, sleeplessness, indigestion, and nausea. Flammable when exposed to heat or flame.

Synonyms: CAS: 8030-30-6 ❖ AROMATIC SOLVENT ❖ BENZIN ❖ COAL TAR NAPHTHA ❖ HI-FLASH NAPHTHAETHYLEN ❖ NAPHTA (DOT) ❖ NAPHTHA DISTILLATE ❖ NAPHTHA PETROLEUM ❖ NAPHTHA, SOLVENT ❖ PETROLEUM BENZIN ❖ PETROLEUM DISTILLATES ❖ PETROLEUM ETHER ❖ PETROLEUM NAPHTHA ❖ PETROLEUM SPIRIT ❖ SKELLY-SOLVE-F ❖ VM&P NAPHTHA

NAPHTHALENE

Products: For moth balls, rug cleaner,

Use: As a germicide and an animal

upholstery cleaners, preservative, and antiseptic; squirrel and small animal repellent. deterrent.

Health Effects: Human poison by swallowing. An eye and skin irritant. Can cause nausea, headache, diaphoresis (perspiration), hematuria (blood in urine), fever, anemia, liver damage, vomiting, convulsions, and coma. Poison by swallowing large doses, breathing, or skin absorption.

Synonyms: CAS: 91-20-3 ✲ TAR CAMPHOR ✲ MOTH BALLS ✲ MOTH FLAKES ✲ NAPHTHALIN ✲ NAPHTHENE ✲ WHITE TAR ✲ CAMPHOR TAR

1-NAPHTHALENESULFONIC ACID

Products: In disinfectant soaps.

Use: As an antibacterial (destroys germs).

Health Effects: Derived from interaction of naphthalene and sulfuric acid. Appropriate label precautions must be observed.

Synonyms: CAS: 85-47-2 ✲ α-NAPHTHALENESULFONIC ACID

NAPHTHENIC ACID

Products: For paint driers, solvents, and detergents.

Use: Various.

Health Effects: Moderately toxic by swallowing. A corrosive material.

Synonyms: CAS: 1338-24-5 ✲ AGENAP ✲ NAPHID ✲ SUNAPTIC ACID B ✲ SUNAPTIC ACID C

β-NAPHTHOL

Products: In pigments, dyes, insecticides, pharmaceuticals, perfumes, and antiseptics.

Use: As coloring agent, preservative, pesticide, and germ killer.

Health Effects: Poison by swallowing. A mutagen (changes inherited characteristics). A skin and eye irritant.

Synonyms: CAS: 135-19-3 ✲ 2-NAPHTHOL ✲ 2-HYDROXYNAPHTHALENE ✲ ISONAPHTHOL ✲ 2-NAPHTHALENOL ✲ DEVELOPER SODIUM ✲ DEVELOPER AMS ✲ β-MONOOXYNAPHTHALENE ✲ β-NAPHTHYL ALCOHOL ✲ β-NAPHTHYL HYDROXIDE

β-NAPHTHYL ETHYL ETHER

Products: In perfumes, soaps, toiletries, and cosmetics.

Use: An odorant or fragrance with orange blossom aroma.

Health Effects: Could cause allergic reaction in susceptible individuals.

Synonyms: NEROLIN

NARCOTIC

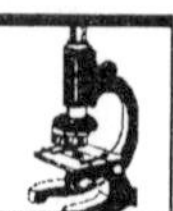
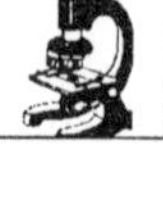

Products: There are natural (morphine, codeine) and synthetic (meperidine, ethadone, and phenazocine) forms.

Use: For medical use.

Health Effects: The sale of narcotics is strictly controlled by law in the U. S. These substances affect sleep (coma) and pain relief, but may result in addiction (a situation where the body can not function normally).

Synonyms: OPIUM ✲ HEROIN ✲ DIONIN

NARINGIN

Products: In beverages.

Use: As sweetener.

Health Effects: Harmless when used for intended purposes.

Synonyms: CAS: 10236-47-2 ✲ NARINGENIN-7-RHAMNOGLUCOSIDE ✲ NARINGENIN-7-RUTINOSIDE ✲ AURANTIIN

NEATSFOOT OIL

Products: For leather and wool items.

Use: As polishing, lubricating, preserving, and finishing agent.

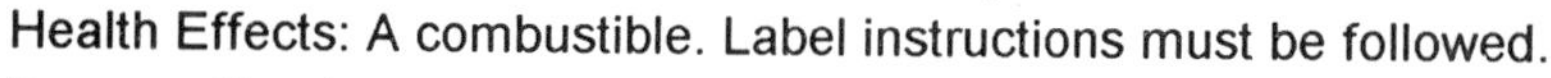

Health Effects: A combustible. Label instructions must be followed.

Synonyms: None known.

NEOPRENE

Products: In adhesives, cements, carpet backings, sealants, gaskets, and adhesive tapes. In both liquid and foam.

Use: For coating, sealing, waterproofing, adhering, materials.

Health Effects: Poison by swallowing. Moderately toxic by breathing. Can cause dermatitis, eye inflammation, corneal necrosis, anemia, hair loss, nervousness, and irritability,

Synonyms: CAS: 126-99-8 ✲ POLYCHLOROPRENE ✲ CHLOROBUTADIENE ✲ CHLOROPRENE

NEOTRIDECANOIC ACID

Products: In lubricants, plasticizers, paint driers, fungicides, and cosmetics.

Use: As an emulsifier (stabilizes and maintains mixes), and as a sequestrant (binds ingredients that affect the final products appearance, flavor, or texture.)

Health Effects: Can cause allergic reactions.

Synonyms: None known.

NICKEL

Products: Coins, batteries, electrical components, jewelry, and alloys.

Use: As an alloy, protective coating, and hydrogenation of vegetable oils.

Health Effects: A confirmed carcinogen. Poison by swallowing. Swallowing of soluble salts causes nausea, vomiting, and diarrhea. A mutagen (changes inherited characteristics). Hypersensitivity to nickel is common and can cause allergic contact dermatitis, asthma, conjunctivitis (inflammation of tissue surrounding eye), inflammatory reactions around nickel-containing medical implants, and prostheses.

Synonyms: CAS: 7440-02-0 ❊ Ni 270 ❊ NICKEL 270 ❊ NICKEL (DUST) ❊ NICKEL SPONGE ❊ Ni 0901-S ❊ Ni 4303T ❊ RANEY ALLOY ❊ RANEY NICKEL

NICOTINE

Products: In tobacco cigarettes, cigars, pipe tobacco, tobacco products, and smokeless tobacco. In pesticides and in veterinary medicine as an external parasiticide (kills parasites).

Use: As an insecticide, fumigant, and stimulant.

Health Effects: An alkoloid (group of nitrogenous organic compounds, mostly used for pain removers such as cocaine, quinine, caffeine) from tobacco. A deadly human poison in concentration. A human teratogen (abnormal fetal development) by swallowing, causes developmental abnormalities of the cardiovascular system. Causes blood pressure effects. Can be absorbed by intact skin.

Synonyms: CAS: 54-11-5 ❊ BLACK LEAF ❊ NICOCIDE ❊ NICOTINE ALKALOID ❊ ORTH N-4 DUST ❊ XL ALL INSECTICIDE

NIGROSINE

Products: In ink, shoe-polish, leather dyes, wood stains, and textile dyes. Also as a shark repellent.

Use: As a coloring agent.

Health Effects: Harmless when used for intended purposes.

Synonyms: None known.

NISIN PREPARATION

Products: For cheese spreads (pasteurized), canned fruits, and vegetables.

Use: As a preservative and antimicrobial agent.

Health Effects: Harmless when used for intended purposes when used within FDA limitations.

Synonyms: CAS: 1414-45-5 ✲ *Streptococcus lactus*

NITRATES

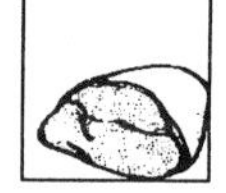

Products: In country hams, bacons, dried meats, and cured sausages.

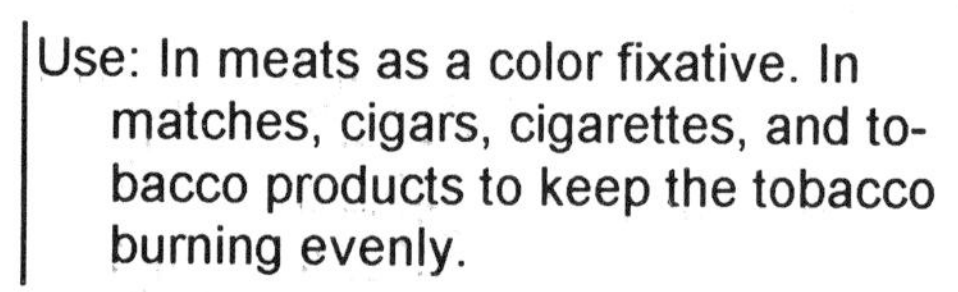

Use: In meats as a color fixative. In matches, cigars, cigarettes, and tobacco products to keep the tobacco burning evenly.

Health Effects: When combined with saliva and secondary amines (food substances), nitrosamines (cancer causing agents) are formed.

Synonyms: CAS: 7757-79-1 ✲ POTASSIUM NITRATE ✲ SODIUM NITRATE

NITRITES

Products: In bacon, meat (cured), meat products, smoked fish, frankfurters, bologna, and poultry products.

Use: Curing agent, color fixative, preserves color, and flavor. Prevents growth of *Botulism* spores.

Health Effects: Large, concentrated amounts of pure substance taken by mouth may produce nausea, vomiting, cyanosis (bluish skin color), collapse, and coma. Repeated small doses cause a fall in blood pressure, rapid pulse, headache, and visual disturbances. They have been implicated in an increased incidence of cancer. FDA approves use to accomplish the effect when used within stated limits. Nitrosamines (cancer causing agents) are formed when nitrites are combined with stomach and food chemicals. FDA states that the addition of Vitamin C reduces the formation of nitrosamines.

Synonyms: POTASSIUM NITRITE ✲ SODIUM NITRITE

NITROBENZENE

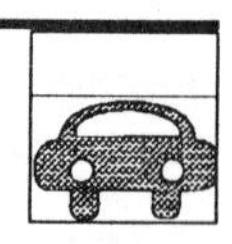

Products: In soaps, cleaning products, metal polishes, shoe polishes, floor, and furniture polishes. Also in oils.

Use: For polishing, cleaning, and lubricating.

Health Effects: Moderately toxic by swallowing and skin contact. Effects on the human body by swallowing include anesthesia, respiratory (breathing) stimulation and vascular (blood vessel) changes. An eye and skin irritant. Absorbed rapidly through the skin. The vapors are hazardous.

Synonyms: CAS: 98-95-3 ❊ OIL OF MIRBANE ❊ ESSENCE OF MIRBANE ❊ OIL OF MYRBANE ❊ MIRBANE OIL

NITROCELLULOSE

Products: In small arms ammunition, explosives, vehicle paint lacquers, printing ink, and celluloid.

Use: Various uses including as a propellant in artillery ammunition

Health Effects: Dangerous fire and explosion risk.

Synonyms: CAS: 9004-70-0 ❊ CELLULOSE NITRATE ❊ NITROCOTTON ❊ GUNCOTTON ❊ PYROXYLIN

NITROGEN

Products: For fruit, poultry, wine, and various food in sealed containers.

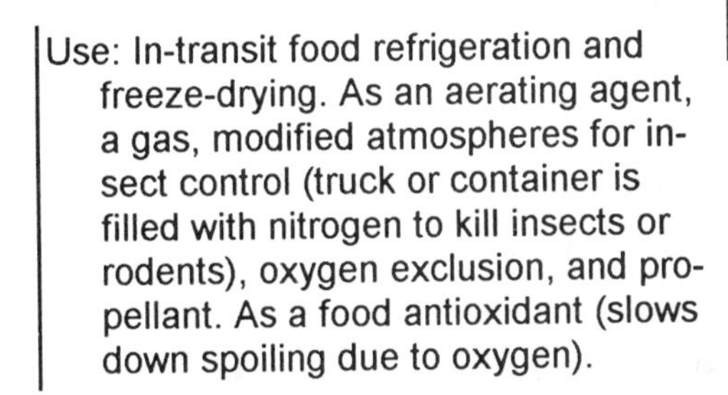

Use: In-transit food refrigeration and freeze-drying. As an aerating agent, a gas, modified atmospheres for insect control (truck or container is filled with nitrogen to kill insects or rodents), oxygen exclusion, and propellant. As a food antioxidant (slows down spoiling due to oxygen).

Health Effects: Low toxicity. In high concentrations it is a simple asphyxiant. The release of nitrogen from solution in the blood, with formation of small bubbles, it is the cause of most of the symptoms and changes found in compressed air illness (Caisson disease, Decompression disease, the Bends). It is a narcotic at high concentration and high pressure. Both the narcotic effects and the bends are hazards of compressed air atmospheres such as found in underwater diving. Nonflammable gas.

Synonyms: CAS: 7727-37-9 ❊ NITROGEN, compressed ❊ NITROGEN, refrigerated liquid ❊ NITROGEN GAS

NITROGEN OXIDE

Products: In simulated dairy products (whipped creams) and in wine.

Use: As an aerating agent, gas, anesthetic in dentistry and surgery; propellant for food and cosmetic products.

Health Effects: Moderately toxic by breathing. Effects on the human body by breathing are general anesthetic, decreased pulse rate without blood pressure fall, and body temperature decrease. An asphyxiant. Moderate explosion hazard.

Synonyms: CAS: 10024-97-2 ✲ DINITROGEN MONOXIDE ✲ FACTITIOUS AIR ✲ HYPONITROUS ACID ANHYDRIDE ✲ LAUGHING GAS ✲ NITROUS OXIDE ✲ NITROUS OXIDE, compressed ✲ NITROUS OXIDE, refrigerated liquid

n-NONYL ALCOHOL

Products: Fruit flavor in beverages, desserts, candies, and ice creams. Fruit odorant in perfume, cosmetics, and toiletries.

Use: A seasoning or scenting ingredient to affect taste of smell of products.

Health Effects: Could cause allergic reaction in susceptible individual.

Synonyms: CAS: 143-08-8 ✲ NONANOL ✲ ALCOHOL C-9 ✲ OCTYL CARBINOL ✲ PELARGONIC ALCOHOL

NONYL PHENOL

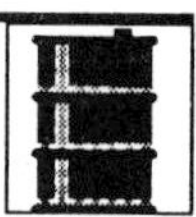

Products: In detergents, emulsifiers, stain removers, oil additives, fungicides, and preservatives, for plastics and rubbers.

Use: Various uses including solvent-type products.

Health Effects: Moderately toxic by swallowing and skin contact.

Synonyms: CAS: 25154-52-3 ✲ ISOMERIC MONOALKYL PHENOL MIXTURE

NORMAL PARAFFIN SOLVENTS

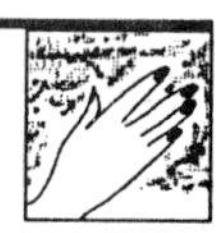

Products: Waterless hand cleaner or solvent formulation.

Use: As cleaning, soil-dissolving, materials.

Health Effects: Harmless when used for intended purposes.

Synonyms: NORPAR

NUTMEG OIL, EAST INDIAN

Products: As fruit, berry, chocolate, spice, and vanilla flavoring in cakes, eggnog, fruit, and puddings.

Use: As taste enhancer or seasoning.

Health Effects: Moderately toxic by swallowing excessive amounts of pure substance. A mutagen (changes inherited characteristics). A skin irritant. GRAS (generally regarded as safe) when used at moderate levels for intended purposes.

Synonyms: CAS: 8008-45-5 ✲ MYRISTICA OIL ✲ NUTMEG OIL ✲ OIL of MYRISTICA ✲ OIL of NUTMEG

Of the 67 pesticides permitted on strawberries, 7 are possible carcinogens.

OCHER

Products: Various colored earthy powders in paint pigments, stains, inks, wall paper pigment, artists' color, cosmetics, and theatrical makeup

Use: As a coloring agent.

Health Effects: Harmless when used for intended purposes.

Synonyms: UMBER ❊ SIENNA ❊ HYDRATED FERRIC OXIDES MIXTURES ❊ CALCINED (BURNT OCHER)

γ-OCTALACTONE

Products: In baked goods, candy, or ice cream.

Use: A peach-flavored additive.

Health Effects: The pure chemical is mildly toxic by swallowing large quantities. A skin irritant. FDA approves use at a moderate level to accomplish the intended effect.

Synonyms: CAS: 104-50-7 ❊ γ-n-BUTYL-γ-BUTYROLACTONE ❊ 5-HYDROXYOCTANOIC ACID LACTONE ❊ OCTANOLIDE-1,4 ❊ TETRAHYDRO-6-PROPYL-2H-PYRAN-2-ONE

1-OCTANAL

Products: In soaps, cosmetics, perfumes, and toiletry products.

Use: An odorant. An ingredient with a strong aromatic scent that affects the aroma of products.

Health Effects: It is mildly toxic by swallowing and skin contact. A skin and eye irritant. FDA approves use at moderate levels to accomplish the intended effect.

Synonyms: CAS: 124-13-0 ❊ ALDEHYDE C-8 ❊ C-8 ALDEHYDE ❊ FEMA No. 2797 ❊ OCTANALDEHYDE ❊ n-OCTYL ALDEHYDE

OCTANE

Products: In fuel and gasoline.

Use: Anti-knocking agent in internal combustion engines.

Health Effects: May act as a simple asphyxiant. A narcotic in high concentrations. Extended skin contact can cause blisters. Brief skin contact causes burning sensation. A dangerous fire and explosion hazard.

Synonyms: CAS: 111-65-9 ❖ OKTAN ❖ OKTANEN ❖ OTTANI

OCTANOIC ACID

Products: In dyes, drugs, perfumes, antiseptics and synthetic flavors. In baked goods, candy (soft), cheese, fats, frozen dairy desserts, gelatins, meat products, oil, packaging materials, puddings, and snack foods.

Use: An additive; an antimicrobial, defoaming, and flavoring agent. A lubricant.

Health Effects: Mildly toxic by swallowing excessive amounts of pure acid. A mutagen (changes inherited characteristics). A skin irritant. Yields irritating vapors which can cause coughing. GRAS (generally regarded as safe) when used within FDA limitations.

Synonyms: CAS: 124-07-2 ❖ C-8 ACID ❖ CAPRYLIC ACID ❖ n-CAPRYLIC ACID ❖ 1-HEPTANECARBOXYLIC ACID ❖ HEXACID 898 ❖ NEO-FAT 8 ❖ OCTIC ACID ❖ n-OCTOIC ACID ❖ n-OCTYLIC ACID

OCTHILINONE

Products: In cooling-tower water, paints, cosmetics, and shampoos.

Use: As mildewcide (destroys mold and mildew), fungicide, and biocide (kills bacteria).

Health Effects: Moderately toxic by swallowing excessive quantities and by skin contact. A severe skin and eye irritant.

Synonyms: CAS: 26530-20-1 ❖ KATHON LP PRESERVATIVE ❖ MICROCHEK ❖ PANCIL ❖ SKANE

OCTYL ALCOHOL

Products: In beverages, candy, gelatin desserts, ice cream, and pudding mixes. In colognes and cosmetics.

Use: As antifoaming, scenting, flavoring agent; an intermediate and solvent.

Health Effects: Moderately toxic by swallowing large quantiites of pure substance. A mutagen (changes inherited characteristics). A skin irritant. FDA approves use only for encapsulating lemon, lime, orange, peppermint, and spearmint oil in limited levels.

Synonyms: CAS: 111-87-5 ❖ ALCOHOL C-8 ❖ ALFOL 8 ❖ CAPRYL ALCOHOL ❖ CAPRYLIC ALCOHOL ❖ DYTOL M-83 ❖ EPAL 8 ❖ HEPTYL CARBINOL ❖ 1-HYDROXYOCTANE ❖ LOROL 20 ❖ OCTANOL ❖ n-OCTANOL ❖

1-OCTANOL ✲ OCTILIN ✲ OCTYL ALCOHOL, NORMAL-PRIMARY ✲ PRIMARY OCTYL ALCOHOL ✲ SIPOL L8

OCTYL ACETATE

Products: In perfumed cosmetics, hair products, skin lotions, and fragrances.

Use: An odorant. An ingredient that affects the aroma of products.

Health Effects: Moderately toxic by swallowing. A skin and eye irritant.

Synonyms: CAS: 103-09-3 ✲ ACETATE C-8 ✲ CAPRYLYL ACETATE ✲ 2 ETHYLHEXANYL ACETATE ✲ 2 ETHYLHEXYL ACETATE

OCTYL GALLATE

Products: In margarine; oleomargarine.

Use: As antioxidant (slows down the spoiling of fats due to oxidation).

Health Effects: Mildly toxic by swallowing great amounts of concentrated chemical. USDA approves use within limitations.

Synonyms: CAS: 1034-01-1

ODORLESS LIGHT PETROLEUM HYDROCARBONS

Products: Beet sugar, eggs, fruits, pickles, vegetables, vinegar, and wine.

Use: As a coating agent, defoamer, float; insecticide formulations component.

Health Effects: Harmless when used for intended purposes. FDA approves use at moderate levels to accomplish the intended effect.

Synonyms: None known.

OIL OF CARDAMOM

Products: In liqueurs, cold medications, cough syrups, sauces, candies, and bakery products.

Use: A seasoning and scenting additive. An ingredient that affects the taste or smell of final product.

Health Effects: Could cause allergic reaction in susceptible individuals.

Synonyms: *ELETTARIA CARDAMOMUM*

OIL of LIME, distilled

Products: In bakery products, beverages

Use: A seasoning and scenting additive.

(nonalcoholic), chewing gum, condiments, confections, gelatin desserts, ice cream products, and puddings.

An ingredient used to affect the taste or smell.

Health Effects: A skin irritant. The pure, concentrated chemical is a possible carcinogen which caused tumors in laboratory animals. A mutagen (changes inherited characteristics). FDA states GRAS (generally regarded as safe) when used within reasonable limits.

Synonyms: CAS: 8008-26-2 ❖ DISTILLED LIME OIL ❖ LIME OIL ❖ LIME OIL, distilled ❖ OILS, LIME

OIL of MACE

Products: In bread, cakes, chocolate pudding, and fruit salad.

Use: As a flavoring agent.

Health Effects: The pure substance is moderately toxic by swallowing excessive amounts. A skin irritant. FDA states GRAS (generally regarded as safe) when used at moderate levels to accomplish the intended effects.

Synonyms: CAS: 8007-12-3 ❖ NCI-C56484 ❖ MACE OIL ❖ OIL of NUTMEG, expressed

OIL of ORANGE

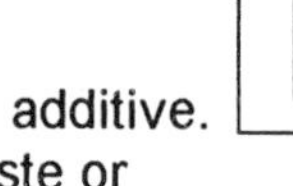

Products: In bakery products, beverages (nonalcoholic), chewing gum, condiments, ice cream products, cough syrup, and cold medications. FDA states GRAS (generally regarded as safe) when used within reasonable limits.

Use: A seasoning and scenting additive. An ingredient that affects taste or smell of product.

Health Effects: Concentrated chemical is a skin irritant.

Synonyms: CAS: 8008-57-9 ❖ NEAT OIL of SWEET ORANGE ❖ OIL of SWEET ORANGE ❖ ORANGE OIL ❖ ORANGE OIL, coldpressed ❖ SWEET ORANGE OIL

OLEIC ACID

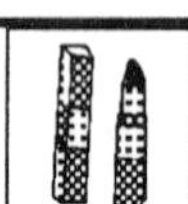

Products: In beet sugar, citrus fruit (fresh), sugar beets, yeast, soaps, ointments, cosmetics, hair wave products, shaving creams, lipstick, shampoos, liquid makeup, nail polishes, polishes, and waterproofing compounds.

Use: As a binding, coating, defoaming, and lubricating agent.

Health Effects: Mildly toxic by swallowing large amounts of concentrated substance. A mutagen (changes inherited characteristics). A skin irritant. A possible carcinogen (causes cancer) that caused tumor growth in laboratory animals. FDA approves use within specific limitations.

Synonyms: CAS: 112-80-1 ✲ CENTURY CD FATTY ACID ✲ EMERSOL ✲ EMERSOL 221 LOW TITER WHITE OLEIC ACID ✲ EMERSOL 220 WHITE OLEIC ACID ✲ GLYCON WO ✲ GROCO 2 ✲ HY-PHI 2102 ✲ INDUSTRENE 105 ✲ K 52 ✲ l'ACIDE OLEIQUE (FRENCH) ✲ METAUPON ✲ NEO-FAT 90-04 ✲ cis-Δ^9-OCTADECENOIC ACID ✲ cis-OCTADEC-9-ENOIC ACID ✲ cis-9-OCTADECENOIC ACID ✲ 9,10-OCTADECENOIC ACID ✲ PAMOLYN ✲ RED OIL ✲ TEGO-OLEIC 130 ✲ VOPCOLENE 27 ✲ WECOLINE OO ✲ WOCHEM No. 320

OLIVE OIL

Products: In soaps, textile soaps, cosmetics, skin creams, medical ointments, liniments, emollient, salad dressings, sardine packing, anchovy packing, and laxatives.

Use: Various uses including lubricating, cooking, and seasoning.

Health Effects: Harmless when used for intended purposes.

Synonyms: CAS: 8001-25-0 ✲ *OLGA EUROPA* ✲ *OLEACEAE*

ORANGE B

Products: In frankfurter and sausage casings.

Use: A coloring ingredient for foods.

Health Effects: FDA approves use when used within limitations.

Synonyms: CAS: 15139-76-1 ✲ 1-(4-SULFOPHENYL)-3-ETHYLCARBOXY-4-(4-SULFONAPHTHYLAZO)-5-HYDROXYPYRAZOLE

ORRIS

Products: A dusting powder, tooth powder, dry shampoos, sachets, dry fragrances, and cosmetics. Also used in ice creams, candies, bakery goods, desserts, and toppings.

Use: A plant root used for fruit/spice flavorings in food and beverages; as an aroma ingredient for toiletries.

Health Effects: Known to cause allergic reactions in susceptible individuals.

Synonyms: ORRIS ROOT OIL ✲ LOVE ROOT ✲ WHITE FLAG

OXALIC ACID

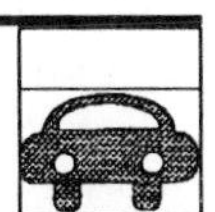

Products: Radiator cleaner, concrete cleaner, rug cleaners, upholstery cleaners, toilet bowl cleaner, ink remover, rust remover, metal polishes, wood cleaners, textile bleaching, and

Use: For printing, dyeing, cleaning, polishing, and color removal.

skin bleaching preparations.

Health Effects: Toxic by breathing and swallowing. A strong irritant.

Synonyms: CAS: 144-62-7 ✲ ETHANEDIOIC ACID

OXYBENZONE

Products: In sunscreen lotions, ointments, sprays, oils, liquids, and creams.

Use: As an UV (ultraviolet) skin protector.

Health Effects: Harmless when used for intended purposes. Could cause an allergic reaction.

Synonyms: CAS: 131-57-7 ✲ 4-METHOXY-2-HYDROXYBENZOPHNONE

OXYMETHUREA

Products: In textiles; antiseptics.

Use: As cotton wrinkleproofing, shrink-proofing, finishing/drying, for wash and wear.

Health Effects: Could cause contact allergic dermatitis in susceptible individuals.

Synonyms: METHURAL ✲ N'N'-BIS(HYDROXYMETHYL)UREA

OXYSTEARIN

Products: For beet sugar, cooking oil, salad oil, vegetable oils, and yeast.

Use: Prevents crystal formation in cooking oils, salad oils; a defoaming agent, and sequestrant (binds constituents that affect the final products flavor or texture).

Health Effects: FDA approves use when used within limitations

Synonyms: None known.

OZOCERITE

Products: In paints, leather polishes, inks, carbon papers, floor polishes, crayons, wax paper, cosmetics, lipstick, rouges, ointments, and waxed cloth. A general purpose wax; sometimes used in place of carnauba and beeswax.

Use: For polishing, protecting, waterproofing, and cosmetic base.

Health Effects: Harmless when used for intended purposes.

Synonyms: MINERAL WAX ✲ FOSSIL WAX ✲ OZOKERITE ✲ CERESIN

OZONE

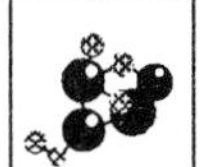

Products: In bottled water. An antimicrobial agent.

Use: For disinfecting, oxidizing, bleaching, deodorizing, and purifying.

Health Effects: A human poison by breathing. Effects on the human body by breathing are: visual field changes, lachrymation (watery eyes), headache, decreased pulse rate with fall in blood pressure, blood pressure decrease, dermatitis, cough, dyspnea (shortness of breath), respiratory stimulation, and other pulmonary changes. A human mutagen (changes inherited characteristics). A skin, eye, upper respiratory system, nose, and throat irritant. Can be a safe water disinfectant in low concentration.

NOTE: Depletion of the ozone layer in the stratosphere which acts as a shield against penetration of UV (ultra-violet) light in the sun's rays is believed to be caused by light-induced chlorofluorocarbon decomposition resulting from increased use of halocarbon aerosol propellants. Their manufacture and use were prohibited in 1979, except for a few specialized items. Chlorofluorocarbon use is being phased out for all home and automobile air conditioners.

Synonyms: CAS: 10028-15-6 ❊ OZON (POLISH) ❊ TRIATOMIC OXYGEN

P

"Coal tar hair dyes are legal, marketing them for use on eyebrows and eyelashes is not legal. It has caused blindness in this application."

FDA Consumer

PALM OIL

Products: In margarine, shortening, soap manufacture, candles, lubricant, cosmetics, ointments, balms, and skin lotions.

Use: As a coating agent, emulsifying agent (stabilizes and maintains mixes), and texturizing agent (improves textures).

Health Effects: GRAS (generally regarded as safe) when used for intended purposes.

Synonyms: None known.

PANTHENOL

Products: Vitamin of the B complex vitamin. In cosmetics, hair shampoos, rinses, emollients (softeners), and dietary supplements. Richest sources are queen bee jelly and liver.

Use: A nutrient. Hair and skin conditioning and softening.

Health Effects: FDA states GRAS (generally regarded as safe) when used for intended purposes.

Synonyms: CAS: 81-13-0 ✲ BEPANTHEN ✲ BEPANTHENE ✲ BEPANTOL ✲ COZYME ✲ DEXPANTHENOL ✲ *d*-(+)-2,4-DIHYDROXY-N-(3-HYDROXYPROPYL)-3,3-DIMETHYLBUTYRAMIDE ✲ D-P-A INJECTION ✲ ILOPAN ✲ MOTILYN ✲ PANADON ✲ PANTHENOL ✲ PANTOL ✲ PANTOTHENOL ✲ PANTOTHENYL ALCOHOL ✲ ZENTINIC

PAPAIN

Products: For meat, poultry, wine and medications. In tobacco, pharmaceuticals, and cosmetics.

Use: For chillproofing (prevents protein haze) of beer; an enzyme (digests protein), for meat tenderizing; cereals (preparation of precooked), processing aid; and tissue softening agent. Used as an anthelmintic (deworming agent), and medication for chronic dyspepsia (indigestion). Derived from papaya.

Health Effects: Effects on the human body by swallowing large, concentrated amounts include changes in structure or function of esophagus. An allergen, can cause allergic reaction. FDA states GRAS (generally regarded as safe) when used within limitations.

Synonyms: CAS: 9001-73-4 ❊ ARBUZ ❊ CAROID ❊ NEMATOLYT ❊ PAPAYOTIN ❊ SUMMETRIN ❊ TROMASIN ❊ VEGETABLE PEPSIN ❊ VELARDON ❊ VERMIZYM

PARADICHLOROBENZENE

Products: In toilet bowl cleaners; moth balls.

Use: As a caustic cleanser; an insect repellent.

Health Effects: A confirmed carcinogen. Moderately toxic to humans by swallowing. Effects on the body by swallowing include changes to the eyes, lungs, thorax, respiration, and decreased motility or constipation. Can cause liver injury. An eye irritant.

Synonyms: CAS: 106-46-7 ❊ PARACHLOROPHENYL CHLORIDE ❊ DICHLORICIDE ❊ DICHLOROBENZENE-PARA ❊ EVOLA ❊ PARA CRYSTALS ❊ SANTA-CHLOR

PARAFFIN

Products: In chewing gum, candles, wax paper, adhesive component, sealants, lubricants, crayons, floor polishes, cold cream, eyebrow pencils, lipliner pencils, lipstick, chewing gum, pharmaceutical base, and wax depilatories.

Use: For coatings, toiletries, lubricating material, masticatory (chewing) substance, food product sealant (for cheeses, cold cuts), tobacco products packaging, water proofing, floor polishing, and glass polishing preparations.

Health Effects: A possible carcinogen that caused tumors in laboratory animals. Many paraffin waxes contain carcinogens. FDA approves use within limitations to produce the desired results.

Synonyms: CAS: 8002-74-2 ❊ PARAFFIN WAX ❊ PARAFFIN WAX FUME

PARAFORMALDEHYDE

Products: In maple syrup, insecticide, disinfectant, contraceptive cream, and bactericide.

Use: Destroys fungus and bacterial growth.

Health Effects: Pure chemical is moderately toxic by swallowing excessive amounts. A severe eye and skin irritant. A mutagen (changes inherited characteristics). FDA approves use when used within strict limitations.

Synonyms: CAS: 30525-89-4 ❊ FLO-MOR ❊ FORMAGENE ❊ PARAFORSN ❊ TRIFORMOL ❊ TRIOXYMETHYLENE

PARAQUAT

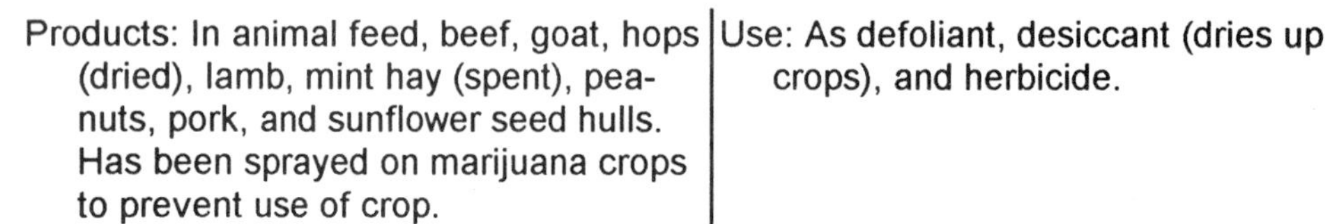

Products: In animal feed, beef, goat, hops (dried), lamb, mint hay (spent), peanuts, pork, and sunflower seed hulls. Has been sprayed on marijuana crops to prevent use of crop.

Use: As defoliant, desiccant (dries up crops), and herbicide.

Health Effects: Poison by swallowing. A mutagen (changes inherited characteristics). Causes ulceration of digestive tract, diarrhea, vomiting, renal (kidney) damage, jaundice (yellowing of skin), edema (fluid accumulation), hemorrhage, fibrosis of lung, and death from anoxia (absence of oxygen) may result. FDA approves use within limitations.

Synonyms: CAS: 4685-14-7 ✼ DIMETHYL VIOLOGEN ✼ GRAMOXONE S ✼ METHYL VIOLOGEN (2+) ✼ PARAQUAT DICATION

PCB

Products: In electrical transformers, components, and systems.

Use: For insulating, heat exchange fluid, and hydraulic fluid.

Health Effects: On EPA Extremely Hazardous Substance list. A confirmed carcinogen (causes cancer). Moderately toxic by swallowing. Causes skin irritation. Exposed persons may suffer nausea, vomiting, weight loss, jaundice (yellow coloration of eyes and skin), edema (swelling from fluid retention), and abdominal (stomach) pain. Where liver damage has been severe the patient may suffer coma or death.

Synonyms: CAS: 1336-36-3 ✼ AROCLOR ✼ CHLOPEN ✼ CHLOREXTOL ✼ CHLORINATED BIPHENYLS ✼ CHLORINATED DIPHENYLS ✼ CLOPHEN ✼ PHENOCHLOR ✼ POLYCHLORBIPHENYL ✼ PYRALENE ✼ PYRANOL ✼ SANTOTHERM

PEANUT OIL

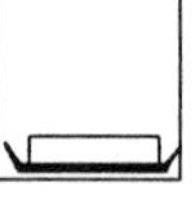

Products: In shortening, salad dressings, mayonnaise, confections, margarine, soap, paint, and as a vehicle for intramuscular injections.

Use: For cooking and seasoning; a suspending agent.

Health Effects: Concentrated substance is a human skin irritant and mild allergen.

Synonyms: CAS: 8002-03-7 ✼ ARACHIS OIL ✼ EARTHNUT OIL ✼ GROUNDNUT OIL ✼ KATCHUNG OIL ✼ PECAN SHELL POWDER

PECTIN

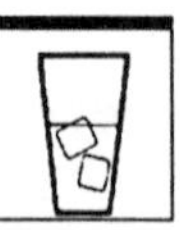

Products: In beverages, jams, jellies, and cosmetics.

Use: As an emulsifier (stabilizes and maintains mixes), gelling agent, stabilizer (to accomplish uniform consistency), and thickening agent.

Health Effects: FDA states GRAS (generally regarded as safe) when used at moderate levels to accomplish the intended effects.

Synonyms: CAS: 9000-69-5

PENTACHLOROPHENOL

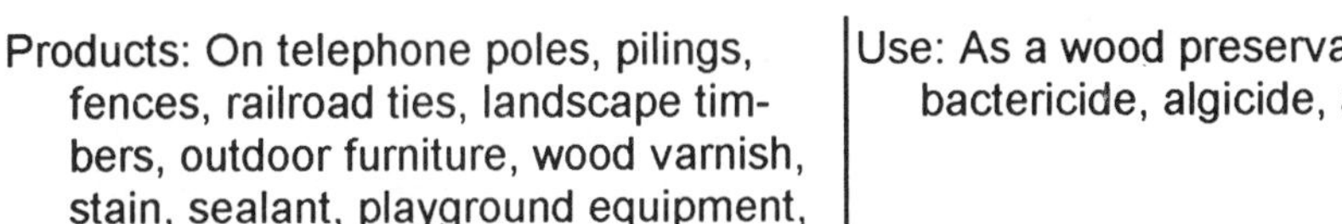

Products: On telephone poles, pilings, fences, railroad ties, landscape timbers, outdoor furniture, wood varnish, stain, sealant, playground equipment, and outdoor wood products. Also in laundry spray starch.

Use: As a wood preservative, fungicide, bactericide, algicide, and herbicide.

Health Effects: Toxic by swallowing, breathing, and skin absorption.

Synonyms: CAS: 87-86-5 ✲ SODIUM PENTACHLOROPHENATE ✲ CHLOROPHEN ✲ DOWCIDE ✲ FUNGIFEN ✲ PENTACHLOR ✲ PENTASOL ✲ PERMAGARD ✲ SANTOPHEN ✲ THOMPSON'S WOOD FIX

PEPPERMINT OIL

Products: In bakery products, beverages (alcoholic), beverages (nonalcoholic), chewing gum, confections, fruit cocktails, gelatin desserts, ice cream, meat, puddings, sauces, and toppings. Also toothpastes, powders, mouth washes, shaving soaps, lotions, over-the-counter antacids, and medications.

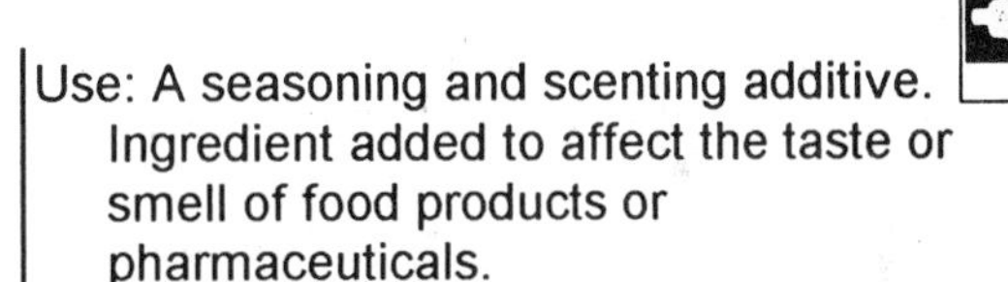

Use: A seasoning and scenting additive. Ingredient added to affect the taste or smell of food products or pharmaceuticals.

Health Effects: Moderately toxic by swallowing excessive quantities of pure substance. An allergen, causes allergic reaction. A mutagen (changes inherited characteristics). FDA states GRAS when used at moderate levels to accomplishes the intended effect.

Synonyms: CAS: 8006-90-4 ✲ PFEFFERMINZ OEL (GERMAN)

PERCHLOROETHYLENE

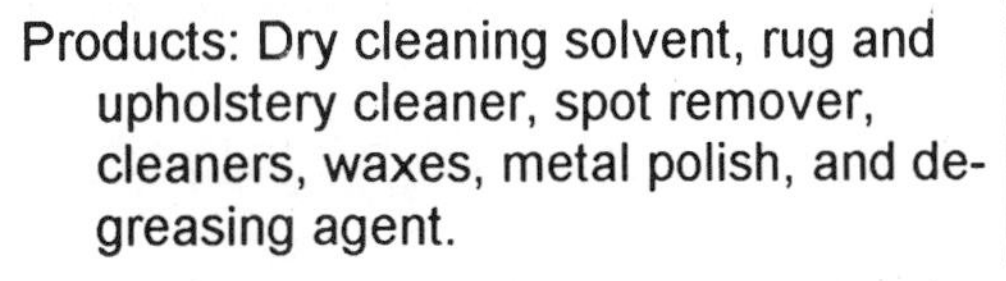

Products: Dry cleaning solvent, rug and upholstery cleaner, spot remover, cleaners, waxes, metal polish, and degreasing agent.

Use: Cleaning agent in 85% of the 30,000 U.S. dry cleaning stores. They emit 92,000 tons of perc into the air each year.

Health Effects: A probable carcinogen (may cause cancer) on EPA list. Moderately toxic to humans by breathing with the following effects: local anesthetic, conjunctiva (eye) irritation, general anesthesia, hallucinations, distorted perceptions (confusion), coma, and pulmonary (lung) changes. A severe eye and skin irritant. The symptoms of acute intoxication from this material are the result of its effects upon the nervous system. Can cause dermatitis, particularly after repeated or prolonged contact with the skin. People who live near, or work at dry cleaners, have been exposed to perc levels hundred of times higher than the acceptable guidelines. Newly dry cleaned clothing should not be stored in children's rooms as they are more sensitive to toxic substances.

Synonyms: CAS: 127-18-4 ✲ ANKILOSTIN ✲ CARBON BICHLORIDE ✲ CARBON DICHLORIDE ✲ DIDAKENE ✲ DOW-PER ✲ ETHYLENE TETRACHLORIDE ✲ FEDAL-UN ✲ NEMA ✲ PERAWIN ✲ PERCHLOR ✲ PERCHLORETHYLENE ✲ PERCLENE ✲ PERCOSOLVE ✲ PERK ✲ PERKLONE ✲ PERSEC ✲ TETLEN ✲ TETRACAP ✲ TETRACHLOROETHENE ✲ 1,1,2,2-TETRACHLOROETHYLENE ✲ TETRALENO ✲ TETRALEX ✲ TETRAVEC ✲ TETROGUER ✲ TETROPIL

PETROLATUM

Products: In bakery products, beet sugar, confectionery, egg white solids, fruits (dehydrated), vegetables (dehydrated), and yeast. Also in ointment bases, pharmaceuticals, cosmetics, rust preventative, modeling clay, and laxatives.

Use: As lubricating, polishing agent, coating (protective), release agent, and sealing agent.

Health Effects: Harmless when used for intended purposes. FDA approves when used within limitations.

Synonyms: WHITE PETROLATUM ✲ YELLOW PETROLATUM ✲ PARAFFIN JELLY ✲ VASOLIMENT ✲ KREMOLINE ✲ VASOLINE ✲ PURELINE

PHENETHYL ALCOHOL

Products: In synthetic rose oil, soaps, flavors, antibacterial, and preservative.

Use: An antimicrobial (destroys bacteria) and odorant. Ingredient to affect taste or smell of products.

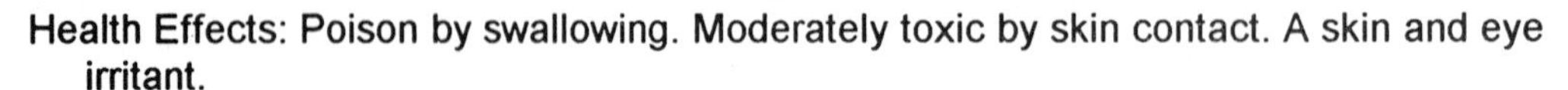

Health Effects: Poison by swallowing. Moderately toxic by skin contact. A skin and eye irritant.

Synonyms: CAS: 60-12-8 ✲ PHENYLETHYL ALCOHOL ✲ 2-PHENYLETHANOL ✲ BENZYL CARBINOL

PHENOL

Products: Topical anesthetic (skin numbing agent). Disinfectant for toilets, outbuildings, floors, drains, and stables.

Use: A germicide, slimicide, and cleansing agent. Causes loss of feeling on skin.

Health Effects: Toxic by swallowing, breathing, and skin absorption. Strong irritant to tissue.

Synonyms: CAS: 108-95-2 ✲ BENZOPHENOL ✲ CRESOLS ✲ XYLENOLS ✲ RESORCINOL ✲ NAPTHOLS ✲ CARBOLIC ACID ✲ PHENYLIC ACID ✲ BENZOPHENOL ✲ HYDROXYBENZENE

PHENOXYACETIC ACID

Products: Corn remover plasters, pads, drops, callus removers, and exfoliants.

Use: To soften and remove hard skin surfaces.

Health Effects: Moderately toxic by swallowing. A mild irritant.

Synonyms: CAS: 122-59-8 ✲ GLYCOLIC ACID PHENOL ETHER ✲ PHENOXYETHANOIC ACID ✲ o-PHENYLGLYCOLIC ACID ✲ PHENYLIUM

p-PHENYLENEDIAMINE

Products: For hair items.

Use: In dyes and coloring products.

Health Effects: An allergen that could cause allergic reaction. May produce eczema and contact dermatitis.

Synonyms: CAS: 106-50-3 ✲ p-DIAMINOBENZENE

PHENYLETHYL ANTHRANILATE

Products: Grape or orange flavor in beverages, synthetic juices, soft drinks, ice creams, ices, candies, and bakery products.

Use: As a flavoring or odorant. An ingredient to affect the taste or smell of product.

Health Effects: Can cause allergic reaction in susceptible individuals.

Synonyms: CAS: 133-18-6 ✲ 2-PHENYLETHYL ANTHRANILATE

PHENYLMERCURIC ACETATE

Products: In paints.

Use: A fungicide, herbicide (destroys plant growth), slimicide, and mildewcide (kills mold and mildew).

Health Effects: Toxic by swallowing, breathing and skin absorption. A strong irritant. On EPA Extremely Hazardous Substance list.

Synonyms: CAS: 62-38-4 ✲ CELMAR ✲ CERESAN ✲ GALLOTOX ✲ KWIKSAN ✲ LEYTOSAN ✲ PAMISAN ✲ SPORKIL

PHENYL SALICYLATE

Products: In adhesives, waxes, polishes, medications, suntan lotions, and creams.

Use: As a preservative, external disinfectant, intestinal antiseptic, and UV (ultra-violet) absorber.

Health Effects: Toxic by swallowing.

Synonyms: CAS: 118-55-8 ✲ SALOL ✲ 2-HYDROXYBENZOIC ACID PHENYL ESTER

PHLOROGLUCINOL

Products: Cut flower preservative; textile dyes; and printing inks.

Use: For foilage preserving and coloring additive.

Health Effects: Mildly toxic by swallowing.

Synonyms: CAS: 108-73-6 ✲ DILOSPAN S ✲ PHLOROGLUCIN ✲ 1,3,5-BENZENETRIOL ✲ 1,3,5-TRIHYDROXYBENZENE

PHOSPHONE ALKYL AMIDE

Products: For 100% cotton fabrics, tents, military uniforms, draperies, canvas, denims, and camping equipment.

Use: As a flame and fire retardant for fabrics.

Health Effects: Harmless when used for intended purposes.

Synonyms: PYROVATEX

PHOSPHORIC ACID

Products: In beverages, cheeses, colas, fats (poultry), lard, margarine, oleomargarine, poultry, root beer, detergents, and shortening. In rust-proofing products and dental cements.

Use: An acid, sequestrant (binds constituents that affect the final products appearance, flavor or texture), increases the preservative effect. The chemical that gives soft drinks their biting taste.

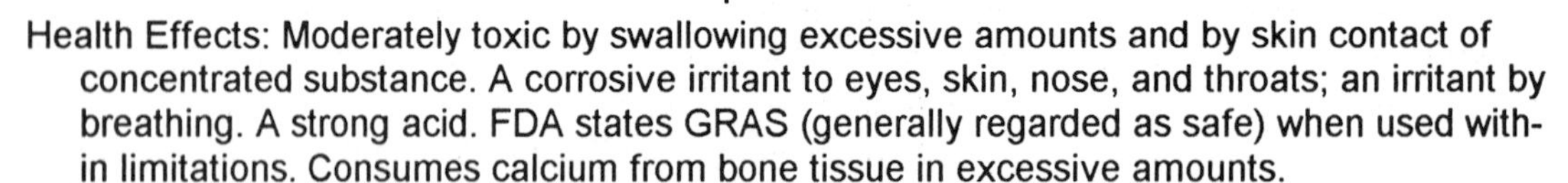

Health Effects: Moderately toxic by swallowing excessive amounts and by skin contact of concentrated substance. A corrosive irritant to eyes, skin, nose, and throats; an irritant by breathing. A strong acid. FDA states GRAS (generally regarded as safe) when used within limitations. Consumes calcium from bone tissue in excessive amounts.

Synonyms: CAS: 7664-38-2 ✲ ACIDE PHOSPHORIQUE (FRENCH) ✲ ACIDO FOSFORICO (ITALIAN) ✲ FOSFORZUUROPLOSSINGEN (DUTCH) ✲ ORTHOPHOSPHORIC ACID ✲ PHOSPHORSAEURELOESUNGEN (GERMAN)

PHOSPHOROUS

Products: *Red:* Safety matches. *White:* A rodenticide (rat and mouse poison). In agricultural fertilizers.

Use: Various.

Health Effects: *Red:* A relatively non-toxic form. *White:* A highly toxic form. Small amounts swallowed may cause bloody diarrhea, severe stomach/intestine irritation, liver damage, circulatory collapse, coma, convulsions, and death.

Synonyms: CAS: 7723-14-0 ❖ PHOSPHOROUS, AMORPHOUS

PHOSPHOROUS, OXYCHLORIDE

Products: A flame-retarding agent for textiles and canvas materials.

Use: For fireproofing.

Health Effects: Toxic by breathing and swallowing. A strong irritant to skin and tissue.

Synonyms: CAS: 10025-87-3 ❖ PHOSPHORYL CHLORIDE

PHOSPHOTUNGSTIC ACID

Products: Anti-static coatings and sprays for textiles. Water-proofing agent for materials.

Use: Various.

Health Effects: A strong irritant to skin and eyes. Some individuals exhibit allergic reactions to these products.

Synonyms: CAS: 12067-99-1 ❖ PHOSPHO-12-TUNGSTIC ACID ❖ PHOSPHWOLFRAMIC ACID ❖ 12-TUNGSTOPHOSPHORIC ACID

PIGMENT BLUE 15

Products: Paints, enamels, inks, lacquers, colored artists chalks, school chalks, and pencils.

Use: A coloring agent.

Health Effects: Harmless when used for intended purposes.

Synonyms: CI No 74160

PIGMENT BLUE 19

Products: Candles, wax products, and decorative items.

Use: A coloring agent.

Health Effects: Harmless when used for intended purposes.

Synonyms: CI No 42750A

PINE NEEDLE OIL

Products: In cleaning materials, deodorizer, air freshner, soaps, toiletries.

Use: As a solvent, disinfectant, odorant, an ingredient that affects the smell of products.

Health Effects: Could cause allergic reaction. Mildly toxic by swallowing. A human skin irritant. FDA permits use at moderate levels to accomplish the intended effect.

Synonyms: CAS: 8000-26-8 ✲ DWARF PINE NEEDLE OIL ✲ KNEE PINE OIL ✲ LATSCHENKIEFEROL ✲ OIL of MOUNTAIN PINE ✲ PINUS MONTANA OIL ✲ PINUS PUMILIO OIL ✲ YARMOR

PINE TAR

Products: Tar soap, shampoos, bath oils, and cough medications.

Use: As a preservative, disinfectant, or deodorant.

Health Effects: Can cause allergic reaction.

Synonyms: PINUS PALUSTRIS ✲ PINACEAE

PIPERONAL

Products: In perfumes, suntan preparations, lipstick, mosquito repellent, and pediculicide (kills lice). In other concentrations it is used as a floral flavoring agent in food products. FDA approves use within limitations.

Use: As a seasoning, insecticide, and odorant.

Health Effects: Moderately toxic by swallowing. Can cause central nervous system depression. A skin irritant.

Synonyms: CAS: 120-57-0 ✲ 3,4-BENZODIOXOLE-5-CARBOXALDEHYDE ✲ 3,4-DIHYDROXYBENZALDEHYDE METHYLENE KETAL ✲ DIOXYMETHYLENE-PROTOCATECHUIC ALDEHYDE ✲ HELIOTROPIN ✲ 3,4-METHYLENE-DIHYDROXYBENZALDEHYDE ✲ 3,4-METHYLENEDIOXYBENZALDEHYDE ✲ PIPERONALDEHYDE ✲ PIPERONYL ALDEHYDE ✲ PROTOCATECHUIC ALDEHYDE METHYLENE ETHER

POLYBUTYLENE

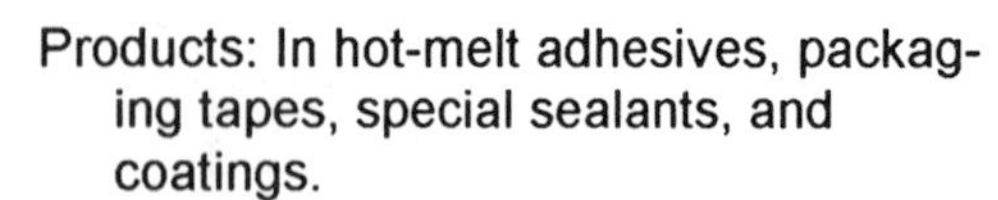

Products: In hot-melt adhesives, packaging tapes, special sealants, and coatings.

Use: As gluing, sealing, sticking, adhering, material.

Health Effects: Items should only be used for intended purposes while closely adhering to label directions.

Synonyms: POLYBUTENE ✲ POLYISOBUTYLENE ✲ POLYISBUTENE

POLYDEXTROSE

Products: In baked goods, baking mixes, (fruit, custard, and pudding-filled pies), cakes, candy (hard), candy (soft), chewing gum, confections, cookies, fillings, frostings, frozen dairy desserts, frozen dairy dessert mixes,

Use: As a bulking agent (a filler), formulation aid, humectant (moisturizer), and texturizing agent.

gelatins, puddings, and salad dressings.

Health Effects: FDA requires special labeling for single servings containing above 15 grams of chemical; "Sensitive individuals may experience a laxative effect from excessive consumption of this product."

Synonyms: CAS: 68424-04-4

POLYETHYLENE

Products: On avocados, bananas, beets, Brazil nuts, chestnuts, chewing gum, coconuts, eggplant, filberts, garlic, grapefruit, hazelnuts, lemons, limes, mangoes, muskmelons, onions, oranges, papaya, peas (in pods), pecans, pineapples, plantain, pumpkin, rutabaga, squash (acorn), sweet potatoes, tangerines, turnips, walnuts, and watermelon. Used in hand lotions.

Use: Masticatory (chewing) substance; coating (protective) for fruits and vegetables to maintain freshness and appearance. For polishing, softening, and fabric finishing.

Health Effects: FDA approves use for intended purposes.

Synonyms: CAS: 9002-88-4 ❖ AGILENE ❖ ALKATHENE ❖ BAKELITE DYNH ❖ DIOTHENE ❖ ETHENE POLYMER ❖ ETHYLENE HOMOPOLYMER ❖ ETHYLENE POLYMERS ❖ HOECHST PA 190 ❖ MICROTHENE ❖ POLYETHYLENE AS ❖ POLYWAX 1000 ❖ TENITE 800

POLYETHYLENE GLYCOL

Products: In beverages (carbonated), citrus fruit (fresh), sodium nitrite coating, sweeteners (nonnutritive), tablets, and vitamin or mineral preparations. FDA approves when used within limitations.

Use: As a binding agent, coating agent, dispersing agent, flavoring additive, lubricant, and plasticizing agent. Also as a softener, humectant, ointments, polish, and base for cosmetics and pharmaceuticals.

Health Effects: Slightly toxic by swallowing excessive amounts of concentrated chemical. An eye irritant.

Synonyms: CAS: 25322-68-3 ❖ CARBOWAX ❖ α-HYDROXY-omega-HYDROXY-POLY(OXY-1,2-ETHANEDIYL) ❖ JEFFOX ❖ LUTROL ❖ PEG ❖ POLY(ETHYLENE OXIDE) ❖ POLY-G SERIES ❖ POLYOX

POLYETHYLENE TEREPHTHALATE

Products: A filament blended with cotton to produce wash and wear fabrics. Blended with wool to produce worsteds

Use: For packaging, textile blending, and various uses.

and suitings. For recording tapes and soft drink bottles. For surgical grafting material.

Health Effects: Harmless when used for intended purposes.

Synonyms: CAS: 25038-59-9 ❖ ALATHON ❖ AMILAR ❖ CELANAR ❖ ESTAR ❖ DACRON ❖ FORTREL ❖ LAVSAN ❖ PEGOTERATE ❖ MYLAR ❖ TERFAN ❖ TERGAL ❖ MELIFORM

POLYISOBUTYLENE

Products: Chewing gum.

Use: Masticatory (chewing) substance in base.

Health Effects: FDA approves use in limited amounts to produce the intended effects.

Synonyms: CAS: 9003-27-4

POLYSORBATE 80

Products: In baked goods, baking mixes, barbecue sauce, beet sugar, chewing gum, confectionery, cottage cheese, custard (frozen), dietary foods (special), dill oil in canned spiced green beans, fillings, food supplements in tablet form, frozen desserts (nonstandardized), fruit sherbet, gelatin dessert mix, ice cream, ice milk, icings, margarine, oil (edible), oil (edible whipped topping), oleomargarine, pickle products, pickles, poultry, shortening, sodium chloride (coarse crystal), toppings, vitamin-mineral preparations with calcium caseinate and fat-soluble vitamins, vitamin-mineral preparations with calcium caseinate in the absence of fat-soluble vitamins, vitamins (fat-soluble), and yeast.

Use: An additive, color fixative, dispersing agent (spreads ingredients), emulsifier (stabilizes and maintains mixes), and stabilizer (used to maintain consistency).

Health Effects: Mildly toxic by swallowing excessive quantities of pure chemical. Human mutation data reported. An eye irritant. FDA permits use within stated limits.

Synonyms: CAS: 9005-65-6 ❖ ARMOTAN PMO-20 ❖ ATLOX 1087 ❖ CRILL 10 ❖ DREWMULSE POE-SMO ❖ DURFAX 80 ❖ ETHOXYLATED SORBITAN MONOOLEATE ❖ GLYCOSPERSE O-20 ❖ LIPOSORB O-20 ❖ MONITAN ❖ MONTANOX 80 ❖ NIKKOL TO ❖ OLOTHORB ❖ POLYOXYETHYLENE SORBITAN MONOOLEATE ❖ POLYOXYETHYLENE SORBITAN OLEATE ❖ POLYSORBAN 80 ❖ PROTASORB O-20 ❖ ROMULGIN O

POLYSTYRENE COPOLYMER EMULSIONS

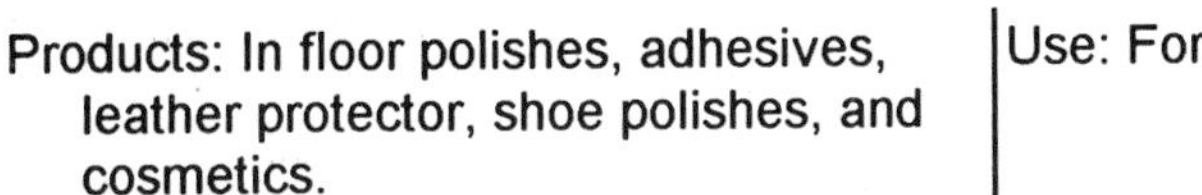

Products: In floor polishes, adhesives, leather protector, shoe polishes, and cosmetics.

Use: For coating and cleaning materials.

Health Effects: An eye, nose, and throat irritant.

Synonyms: URETHANE PREPOLYMERS

POLYURETHANE

Products: In sealants, caulking, mortars, and adhesives. Consumer products come in both rigid and flexible form. The variety of products range from spandex fibers, to cigarette filters, to marine flotation devices.

Use: Various.

Health Effects: Produces toxic fumes upon ignition.

Synonyms: CAS: 9009-54-5 ❖ POLYFOAM PLASTIC ❖ POLYFOAM SPONGE ❖ POLYURETHANE ESTER FOAM ❖ POLYURETHANE SPONGE

POLYVINYL ALCOHOL

Products: In cosmetics, cements, mortars, inks, and greaseproofing.

Use: An emulsifier (stabilizes and maintains mixes), thickener, coatings, binder, and stabilizer (used to keep a uniform consistency).

Health Effects: Harmless when used for intended purposes.

Synonyms: CAS: 9002-89-5 ❖ PVA ❖ PVOH ❖ ELVANOL ❖ ETHENOL HOMOPOLYMER ❖ VINYL ALCOHOL POLYMER

POLYVINYLPYRROLIDNONE HOMOPOLYMER

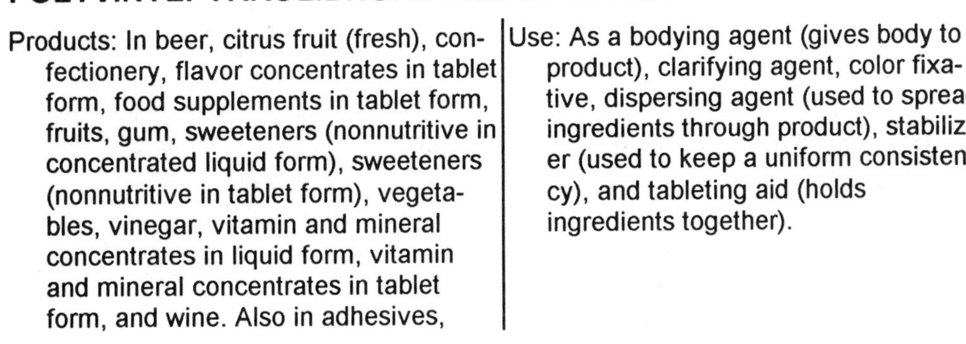

Products: In beer, citrus fruit (fresh), confectionery, flavor concentrates in tablet form, food supplements in tablet form, fruits, gum, sweeteners (nonnutritive in concentrated liquid form), sweeteners (nonnutritive in tablet form), vegetables, vinegar, vitamin and mineral concentrates in liquid form, vitamin and mineral concentrates in tablet form, and wine. Also in adhesives,

Use: As a bodying agent (gives body to product), clarifying agent, color fixative, dispersing agent (used to spread ingredients through product), stabilizer (used to keep a uniform consistency), and tableting aid (holds ingredients together).

cosmetics, shampoos, hand creams, skin lotions, dentifrices, and hair sprays.

Health Effects: FDA approves use in food within limitations.

Synonyms: CAS: 9003-39-8 ❊ ALBIGEN A ❊ ALDACOL Q ❊ BOLINAN ❊ 1-ETHENYL-2-PYRROLIDINONE HOMOPOLYMER ❊ 1-ETHENYL-2-PYRROLIDINONE POLYMERS ❊ HEMODESIS ❊ HEMODEZ ❊ KOLLIDON ❊ LUVISKOL ❊ NEOCOMPENSAN ❊ PERAGAL ST ❊ PERISTON ❊ PLASDONE ❊ POLYCLAR L ❊ POLY(1-(2-OXO-1-PYRROLIDINYL)ETHYLENE) ❊ POLYVIDONE ❊ POLY(n-VINYLBUTYROLACTAM) ❊ PROTAGENT ❊ PVP ❊ SUBTOSAN ❊ VINISIL ❊ N-VINYLBUTYROLACTAM POLYMER ❊ N-VINYLPYRROLIDONE POLYMER

POTASSIUM ACID TARTRATE

Products: In baked goods, candy (hard), candy (soft), confections, crackers, frostings, gelatins, jams, jellies, margarine, oleomargarine, puddings, and wine (grape).

Use: An acid, anticaking agent, antimicrobial agent (destroys germs), humectant (maintains moisture content), leavening agent (aids in mixing ingredients), processing aid, stabilizer (used to maintain uniform consistency), and thickening agent.

Health Effects: FDA permits use within limitations.

Synonyms: CAS: 868-14-4 ❊ CREAM of TARTER ❊ POTASSIUM BITARTRATE

POTASSIUM ALGINATE

Products: In confections, frostings, fruit juices, fruits (processed), gelatins, and puddings.

Use: An emulsifier (stabilizes and maintains mixes), stabilizer (keeps a uniform consistency), and thickening agent.

Health Effects: FDA states GRAS (generally regarded as safe) and approves use within limitations.

Synonyms: CAS: 9005-36-1 ❊ ALGIN

POTASSIUM ARSENATE

Products: Fly paper, insecticidal preparations, leather tanning, and printing textiles.

Use: For pesticides, preservatives, fabric coloring and inking.

Health Effects: Very toxic, as are all arsenic compounds. Confirmed human carcinogen (causes cancer). Can cause a variety of skin problems including itching, skin pigmentation changes, and skin cancers. A possible mutagen (changes inherited characteristics).

Synonyms: CAS: 7784-41-0 ❊ MACQUER'S SALT ❊ MONOPOTASSIUM ARSENATE ❊ POTASSIUM ACID ARSENATE ❊ POTASSIUM DIHYDROGEN ARSENATE ❊ MONOPOTASSIUM DIHYDROGEN ARSENATE

POTASSIUM BENZOATE

Products: In margarine, oleomargarine, and wine.

Use: As a preservative.

Health Effects: Harmless when used for intended purposes. USDA approves use within limitations.

Synonyms: CAS: 582-25-2

POTASSIUM BICARBONATE

Products: In baked goods, margarine, soft drinks, liquid detergents, and oleomargarine. As baking powder and antacid.

Use: An alkali, formulation aid, leavening agent, nutrient supplement, and processing aid.

Health Effects: FDA and USDA permit use within limitations.

Synonyms: CAS: 298-14-6 ❊ POTASSIUM ACID CARBONATE ❊ BAKING SODA

POTASSIUM BROMATE

Products: In baked goods, beverages (fermented malt), confectionery products, permanent wave solutions, tooth pastes, and mouth washes..

Use: As dough conditioner, maturing agent (speeds up the ageing process in order to make more manageable dough); as an antiseptic, and astringent.

Health Effects: A poison by swallowing excessive quantites of pure chemical. An irritant to skin, eyes, nose, and throats. FDA approves use within limitations.

Synonyms: CAS: 7758-01-2 ❊ BROMIC ACID, POTASSIUM SALT

POTASSIUM CARBONATE (2:1)

Products: In margarine, oleomargarine, soups, pigments, printing inks, soft soaps, and liquid shampoos.

Use: Various.

Health Effects: A poison by swallowing concentrated amounts. A strong caustic and very irritating.

Synonyms: CAS: 584-08-7 ❊ PEARL ASH ❊ POTASH ❊ SALT OF TARTAR ❊ CARBONIC ACID, DIPOTASSIUM SALT ❊ K-GRAN

POTASSIUM CHLORATE

Products: In matches, soap, mouthwash, toothpaste, permanent wave lotions, and bleaches.

Use: As disinfectants, astringents, fireworks, and explosives.

Health Effects: Moderately toxic. A gastrointestinal tract and kidney irritant. Can cause dermatitis.

Synonyms: CAS: 3811-04-9 ✲ BERTHOLETT SALT ✲ CHLORATE OF POTASH ✲ SALT OF TARTAR ✲ PEARL ASH ✲ POTASH OF CHLORATE ✲ POTCRATE

POTASSIUM CHLORIDE

Products: In fertilizer, jelly (artificially sweetened), meat (raw cuts), poultry (raw cuts), and preserves (artificially sweetened).

Use: As a dietary supplement, flavor enhancer, flavoring agent, gelling agent, nutrient, salt substitute, tissue softening agent, and yeast food.

Health Effects: FDA states GRAS (generally regarded as safe) when used within specific limitations. However, in pure form it is a human poison by swallowing large amounts. Effects on the human body by swallowing are nausea, blood clotting changes, and cardiac arrhythmias. An eye irritant. A mutagen (changes inherited characteristics).

Synonyms: CAS: 7447-40-7 ✲ CHLOROPOTASSURIL ✲ DIPOTASSIUM DICHLORIDE ✲ EMPLETS POTASSIUM CHLORIDE ✲ ENSEAL ✲ KALITABS ✲ KAOCHLOR ✲ KAON-CI ✲ KAY CIEL ✲ K-LOR ✲ KLOTRIX ✲ K-PRENDE-DOME ✲ PFIKLOR ✲ POTASSIUM MONOCHLORIDE ✲ POTAVESCENT ✲ REKAWAN ✲ SLOW-K ✲ TRIPOTASSIUM TRICHLORIDE

POTASSIUM CITRATE

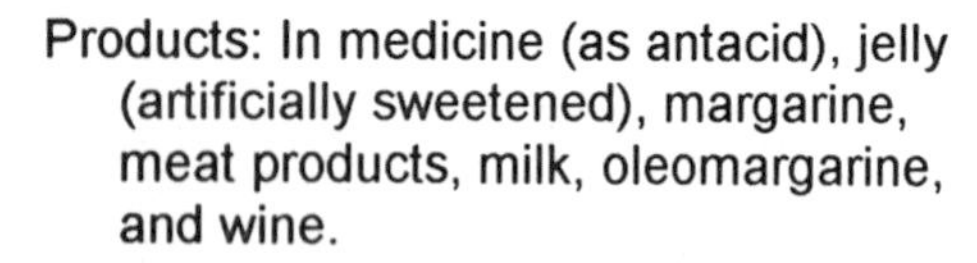

Products: In medicine (as antacid), jelly (artificially sweetened), margarine, meat products, milk, oleomargarine, and wine.

Use: Miscellaneous and general-purpose buffer, sequestrant (binds ingredients that affect the final product appearance, flavor, or texture).

Health Effects: FDA states GRAS (generally regarded as safe) when used within stated limitations.

Synonyms: CAS: 866-84-2 ✲ CITRIC ACID, TRIPOTASSIUM SALT ✲ TRIPOTASSIUM CITRATE MONOHYDRATE

POTASSIUM DICHROMATE

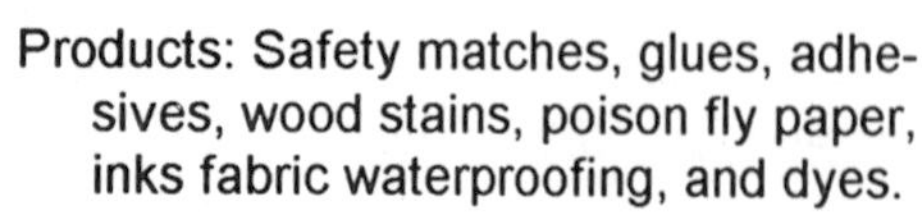

Products: Safety matches, glues, adhesives, wood stains, poison fly paper, inks fabric waterproofing, and dyes.

Use: Various.

Health Effects: Toxic by swallowing and breathing.

Synonyms: CAS: 7778-50-9 ✲ POTASSIUM BICHROMATE ✲ RED POTASSIUM CHROMATE

POTASSIUM HYDROXIDE

Products: In soap, bleach, drain cleaners, liquid fertilizers, oven cleaners, paint remover, varnish removers, cosmetic cuticle remover, shaving lotions, hand creams, and facial blushes.

Use: Various.

Health Effects: Toxic by swallowing and breathing. A strong caustic which is corrosive to tissue in concentrated form. Consumer products contain dilute concentrations.

Synonyms: CAS: 1310-58-3 ✲ CAUSTIC POTASH ✲ POTASSIUM HYDRATE ✲ LYE

POTASSIUM LACTATE

Products: On meat and poultry products, except infant foods and infant formulas.

Use: As a flavor enhancer, flavoring agent, and moisturizing agent.

Health Effects: Harmless when used for intended purposes and within approved limits.

Synonyms: CAS: 996-31-6

POTASSIUM LAURATE

Products: Liquid soaps and shampoos.

Use: An emulsifier (stabilizes and maintains mixes), and a common base.

Health Effects: Harmless when used for intended purposes.

Synonyms: None found.

POTASSIUM PERMANGANATE

Products: For topical antibacterials, deodorizer, bleach, and water purification.

Use: In production of drugs of abuse, disinfectant, and to bleach stone-washed blue jeans.

Health Effects: A human poison by swallowing. Effects on the human body by swallowing are dyspnea (shortness of breath), nausea and gastrointestinal (stomach and intestines) effects. A strong irritant.

Synonyms: CAS: 7722-64-7 ✲ CAIROX ✲ CHAMELEON MINERAL ✲ CONDY'S CRYSTALS

POTASSIUM PEROXYMONOSULFATE

Products: In dry laundry bleaches,

Use: Various.

detergents, washing compounds, cleansers, scouring powders, plastic dishware cleaner, metal cleaners, permanent wave neutralizers, and antiseptics.

Health Effects: Harmless when used for intended purposes while adhering to label directions.

Synonyms: OXONE

POTASSIUM SORBATE

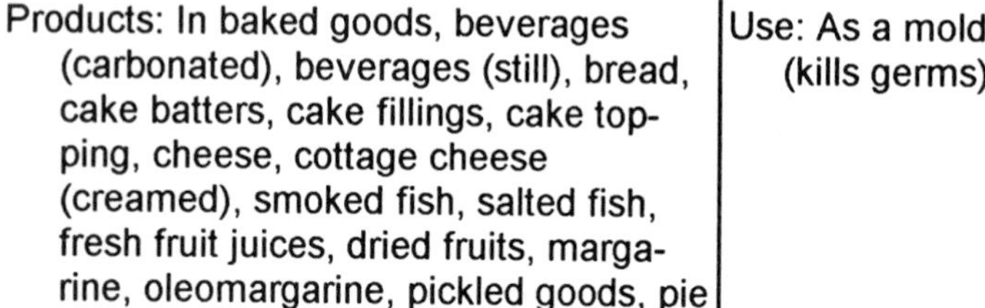

Products: In baked goods, beverages (carbonated), beverages (still), bread, cake batters, cake fillings, cake topping, cheese, cottage cheese (creamed), smoked fish, salted fish, fresh fruit juices, dried fruits, margarine, oleomargarine, pickled goods, pie crusts, pie fillings, salad dressings, salads (fresh), sausage (dry), sea food cocktail, syrups (chocolate dairy), and wine.

Use: As a mold retardant, bacteriostat (kills germs), and preservative.

Health Effects: Mildly toxic by swallowing excessive amounts of pure substance. A possible mutagen (changes inherited characteristics). FDA states GRAS (generally regarded as safe) when used within limitations.

Synonyms: CAS: 590-00-1 ✲ 2,4-HEXADIENOIC ACID POTASSIUM SALT ✲ SORBIC ACID, POTASSIUM SALT ✲ SORBISTAT-K ✲ SORBISTAT-POTASSIUM

POTASSIUM STEARATE

Products: Fabric softener, chewing gum, and packaging materials.

Use: As an anticaking agent, binder, emulsifier (stabilizes and maintains mixes), and stabilizer (used to keep a uniform consistency).

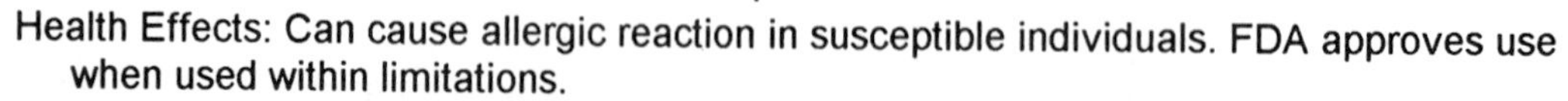

Health Effects: Can cause allergic reaction in susceptible individuals. FDA approves use when used within limitations.

Synonyms: CAS: 593-29-3 ✲ STEARIC ACID POTASSIUM SALT.

POTASSIUM SULFITE

Products: In food and wine.

Use: As a preservative.

Health Effects: GRAS (generally regarded as safe).

Synonyms: CAS: 10117-38-1 ✲ SULFUROUS ACID, DIPOTASSIUM SALT

POTASSIUM TETROXALATE

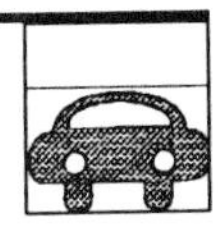

Products: In spot remover, rust remover, ink remover, and metal polishes.

Use: As a cleaner and polish.

Health Effects: Products are harmless when used according to label directions.

Synonyms: CAS: 127-96-8 ✲ POTASSIUM QUADROXALATE ✲ SAL ACETOSELLA ✲ SALT OF SORREL

POTASSIUM UNDECYLENATE

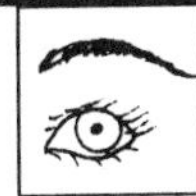

Products: In cosmetics, skin lotions, facial makeup, mascara, eye shadow, and eye liner pencils. In medicinal pharmaceutical items such as skin ointments, creams, and lotions.

Use: As bacteriostat (kills bacteria) and fungistat (kills fungus).

Health Effects: Toxic in high concentrations.

Synonyms: None found.

PROPANE

Products: For cosmetics, shaving creams, foamed foods, cigarette lighters, cooking ranges, and sprayed foods.

Use: An aerating agent, gas refrigerant, and spray propellant.

Health Effects: Central nervous system effects at high concentrations. An asphyxiant. Flammable gas. GRAS (generally regarded as safe) by FDA.

Synonyms: CAS: 74-98-6 ✲ DIMETHYLMETHANE ✲ PROPYL HYDRIDE

PROPIONIC ACID

Products: In perfumes, artificial fruit flavors, breads, and grains.

Use: Antimicrobial agent; flavoring agent; mold inhibitor in bread; preservative, and emulsifying agent (stabilizes and maintains mixes).

Health Effects: In excessive quantities it is moderately toxic by swallowing and skin contact. A corrosive irritant to eyes, skin, nose, and throats. FDA states GRAS (generally regarded as safe)

Synonyms: CAS: 79-09-4 ✲ CARBOXYETHANE ✲ ETHANECARBOXYLIC ACID ✲ ETHYLFORMIC ACID ✲ METACETONIC ACID ✲ METHYL ACETIC ACID ✲ PROPANOIC ACID ✲ PROPIONIC ACID, solution containing not less than 80% acid ❖ PROPIONIC ACID GRAIN PRESERVER ❖ PROZOIN ❖ PSEUDOACETIC ACID ❖ SENTRY GRAIN PRESERVER ❖ TENOX P GRAIN PRESERVATIVE

PROPYLENE GLYCOL ALGINATE

Products: In baked goods, beer, cheese, citrus fruit (raw), condiments, confections, dairy desserts (frozen), fats, flavorings, frostings, gelatins, gravies, ices (fruit and water), jams, jellies, oil, puddings, relishes, salad dressings, sauces (sweet), seasonings, and syrups.

Use: As emulsifier (stabilizes and maintains mixes), stabilizer (used to keep a uniform consistency), and as a thickening agent (for adding body and thickness to the texture).

Health Effects: Mildly toxic by swallowing excessive amounts of pure substance. FDA permits use within limitations.

Synonyms: CAS: 9005-37-2 ❖ HYDROXY PROPYL ALGINATE ❖ KELCOLOID

PROPYLENE GLYCOL PHENYL ETHER

Products: In soaps and perfumes.

Use: A bactericide and fixative for fragrance and cleaning products.

Health Effects: Harmless when used for intended purposes.

Synonyms: None found.

PROPYLPARABEN

Products: In food and pharmaceutical products.

Use: As a preservative, fungicide, and mold preventer. An antimicrobial (kills germs).

Health Effects: Approved by FDA. Mildly toxic by swallowing excessive quantities. An allergen. Could cause allergic reaction.

Synonyms: CAS: 94-13-3 ❖ PROPYLPARASEPT ❖ PARABEN ❖ PARASEPT ❖ ASEPTOFORM ❖ BETACIDE ❖ PRESERVAL ❖ p-HYDROXYPROPYL BENZOATE

PROPYLENE GLYCOL MONO- and DIESTERS

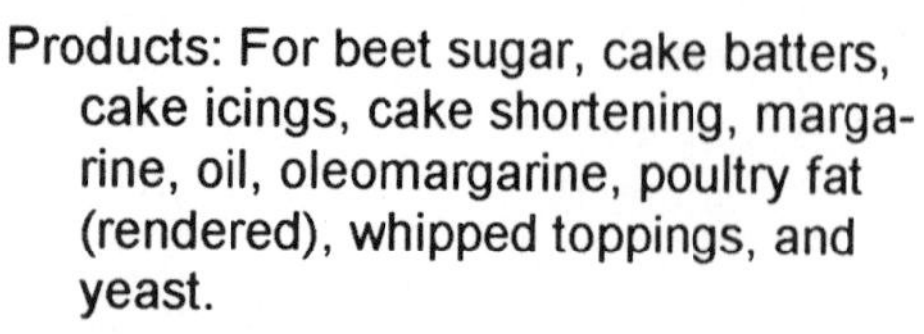

Products: For beet sugar, cake batters, cake icings, cake shortening, margarine, oil, oleomargarine, poultry fat (rendered), whipped toppings, and yeast.

Use: As an emulsifier (stabilizes and maintains mixes) and stabilizer (used to keep a uniform consistency.

Health Effects: FDA and USDA permits use within limitations.

Synonyms: PROPYLENE GLYCOL MONO- and DIESTERS of FATTY ACIDS ❊ PROPYLENE GLYCOL MONOSTEARATE

PUMICE

Products: In cleansers, scouring agents; fireproofing and insulating compounds; cosmetics for removing rough skin; heavy-duty hand soaps; facial cleansing and acne compounds; tooth polishes, and denture powders. Pencil erasers are composed of synthetic rubber and pumice. It is the pumice that erases, not the rubber.

Use: Various abrasive purposes.

Health Effects: Harmless when used for intended purposes.

Synonyms: SEISMOTITE

PYRETHRIN

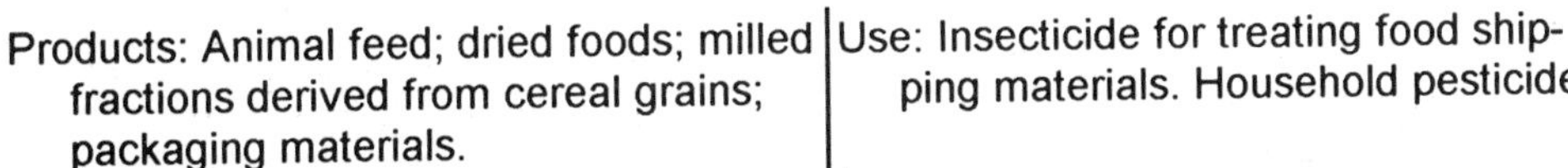

Products: Animal feed; dried foods; milled fractions derived from cereal grains; packaging materials.

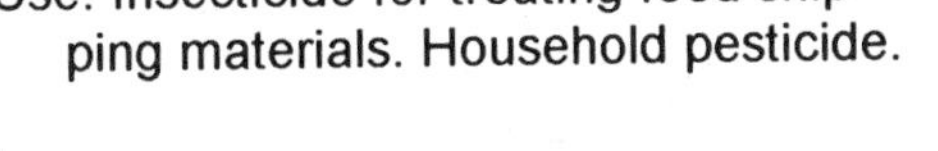

Use: Insecticide for treating food shipping materials. Household pesticide.

Health Effects: Moderately toxic by swallowing pure chemical. Can cause allergic reactions. May cause nausea, vomiting, headache and other effects. FDA permits use within limitations.

Synonyms: CAS: 97-11-0 ❊ 2-CYCLOPENTENYL-4-HYDROXY-3-METHYL-2-CYCLOPENTEN-1-ONE CHRYSANTHEMATE ❊ 3-(2-CYCLOPENTEN-1-YL)-2-METHYL-4-OXO-2-CYCLOPENTEN-1-YL CHRYSANTHEMUMATE ❊ 3-(2-CYCLOPENTENYL)-2-METHYL-4-OXO-2-CYCLOPENTENYL CHRYSANTHEMUMMONOCARBOXYLATE ❊ CYCLOPENTENYLRETHONYL CHRYSANTHEMATE

PYRIDOXOL HYDROCHLORIDE

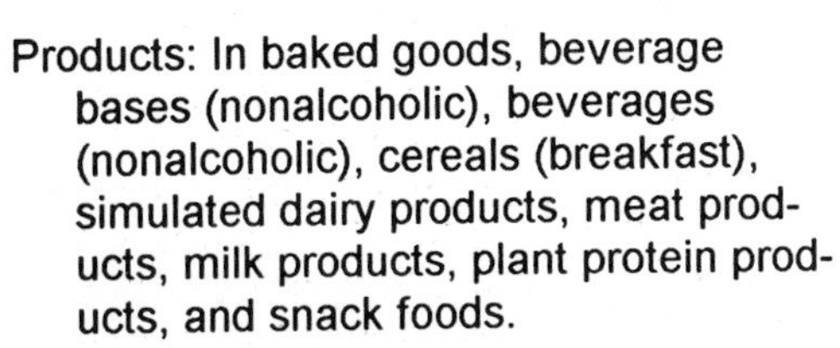

Products: In baked goods, beverage bases (nonalcoholic), beverages (nonalcoholic), cereals (breakfast), simulated dairy products, meat products, milk products, plant protein products, and snack foods.

Use: As a dietary supplement and nutrient.

Health Effects: Concentrated substance is moderately toxic by swallowing large quantities. A mutagen (changes inherited characteristics). FDA states GRAS (generally regarded as safe) when used at moderate levels to accomplish the intended effects.

Synonyms: CAS: 58-56-0 ❊ ADERMINE HYDROCHLORIDE ❊ BECILAN ❊ BENADON ❊ CAMPOVITON 6 ❊ HEBABIONE HYDROCHLORIDE ❊ HEXABETALIN ❊ HEXAVIBEX ❊ HEXERMIN ❊ HEXOBION ❊

3-HYDROXY-4,5-DIMETHYLOL-α-PICOLINE HYDROCHLORIDE ✲ 5-HYDROXY-6-METHYL-3,4-PYRIDINEDICARBINOL HYDROCHLORIDE ✲ 5-HYDROXY-6-METHYL-3,4-PYRIDINEDIMETHANOL HYDROCHLORIDE ✲ 2-METHYL-3-HYDROXY-4,5-BIS(HYDROXYMETHYL)PYRIDINE HYDROCHLORIDE ✲ PYRIDIPCA ✲ PYRIDOXINE HYDROCHLORIDE (FCC) ✲ PYRIDOXINIUM CHLORIDE ✲ PYRIDOXINUM HYDROCHLORICUM (HUNGARIAN) ✲ VITAMIN B6-HYDROCHLORIDE

> "75% of all housing units built before 1980 have lead paint somewhere."
> *Department of Housing and Urban Development*

QUERCITIN

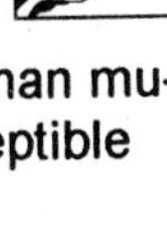

Products: In hair products.

Use: As a dye or coloring agent.

Health Effects: Poison by swallowing. A possible carcinogen (causes cancer). A human mutagen (changes inherited characteristics). Could cause allergic reactions in susceptible individuals.

Synonyms: CAS: 117-39-5 ❊ C.I. NATURAL RED ❊ C.I. NATURAL YELLOW ❊ MELETIN ❊ QUERCETOL ❊ QUERTINE ❊ SOPHORETIN ❊ XANTHAURINE ❊ CYANDIELONON

QUANTERNARY AMMONIUM SALT

Products: In cosmetics, after shave lotions, deodorants, hair colorings, hair curling products, dandruff removers, and detergents.

Use: As a disinfectant, cleanser, sterilizer, fungicide/mildew control, and others.

Health Effects: A poison by swallowing. A skin and severe eye irritant.

Synonyms: CAS: 8001-54-5 ❊ BENZALKONIUM CHLORIDE ❊ DRAPOLENE ❊ PHENEENE GERMICIDAL SOLUTION ❊ ZEPHIRAN CHLORIDE

QUILLAJA EXTRACT

Products: In shampoo, bubble baths, soap products, detergents, soap substitutes; in mineral water industry. In root beer, beverages, spices, candies, and dessert ices.

Use: As a foam producer in cleaning products. As the foam producer in fire extinguishers. As a flavoring ingredient.

Health Effects: Harmless when used for intended purposes.

Synonyms: SOAP BARK ❊ QUILLAY BARK ❊ PANAMA BARK ❊ CHINA BARK ❊ MURILLO BARK

QUININE

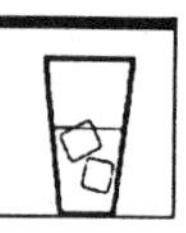

Products: In tonics, medications (antimalarial), beverages, and cosmetics.

Use: As pharmaceutical and flavoring agent.

Health Effects: Effects on the human body by swallowing excessive amounts of concentrated substances are visual changes, tinnitus (ringing in the ears), nausea or vomiting. In concentration it can cause mutagenic (changes inherited characteristics), and teratogenic (abnormal fetal development) effects. A skin irritant particularly to barbers and beauticians who work with quinine tonics.

Synonyms: CAS: 130-95-0 ✻ 6-METHOXYCINCHONINE

QUINOLINE YELLOW

Products: In drugs, cosmetics, facial makeup, and skin creams.

Use: As a colorant.

Health Effects: Synthetic dye approved for use by FDA except in eye area.

Synonyms: C.I. ACID YELLOW ✻ D&C YELLOW NO. 10 ✻ ACID YELLOW 3 ✻ FOOD YELLOW 13

R

"There is no such thing as a nonallergenic toiletry or makeup product."
Color Me Beautiful/Carole Jackson

RADON

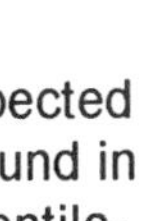

Products: Radioactive gas of radium and uranium decay.

Use: In medical treatment and chemical research.

Health Effects: A common air contaminant. Accumulation of the gas in homes is suspected to be the second leading cause of lung cancer in the U. S. This accumulation is found in well-insulated buildings located over land which has concentrations of uranium. Ventilation prevents accumulation.

Synonyms: CAS: 10043-92-2 ✲ Rn

RAPESEED OIL

Products: In cake mixes, fats (edible), peanut butter, salad dressings, margarines, and shortening. Also in soft hand soaps.

Use: An emulsifier (stabilizes and maintains mixes), stabilizer (used to keep uniform consistency), and as a thickening agent. A food additive, lubricant, rubber substitute, and for oiling woolens.

Health Effects: Harmless when used for intended purposes. FDA states GRAS (generally regarded as safe) when used within limitations.

Synonyms: ✲ COLZA OIL ✲ FULLY HYDROGENATED RAPESEED OIL ✲ LOW ERUCIC ACID RAPESEED OIL ✲ RAPE SEED OIL ✲ SUPERGLYCERINATED FULLY HYDROGENATED RAPESEED OIL

RENNET

Products: In fillings, frozen dairy desserts, gelatins, loaves (nonspecific), milk products, poultry, puddings, sausage, sausage (imitation), soups, and stews.

Use: As a binder, enzyme, extender, processing aid, stabilizer (keeps a uniform consistency), and thickening agent.

Health Effects: FDA states GRAS (generally regarded as safe).

Synonyms: CAS: 9001-98-3 ✲ BOVINE RENNET

RESORCINOL

Products: In adhesives, cosmetics, hair coloring, antidandruff shampoos, lipstick, and hair care products.

Use: As an antiseptic, antipruritic (antiitching agent), preservative, dye, and astringent.

Health Effects: A possible allergen. Could be an irritant to eyes, nose, throat, and skin.

Synonyms: CAS: 108-46-3 ✲ RESORCIN ✲ m-DIHYDROXYBENZENE ✲ 3-HYDROXYPHENOL

RESVERATROL

Products: In red wines and grape juice.

Use: Believed to be the ingredient in grapes that lowers artery-clogging low-density lipoprotein (LDL) cholesterol, while elevating good high-density lipoprotein (HDL) cholesterol. The HDL helps flush coronary arteries of fatty deposits.

Health Effects: Harmless when used in moderate levels.

Synonyms: None found.

RHODINOL

Products: In food, beverages, and cosmetics. Produces a synthetic rose-like aroma.

Use: An ingredient to affect the taste or smell of products.

Health Effects: Could cause allergic reaction in susceptible individuals.

Synonyms: CAS: 6812-78-8 ✲ *l*-CITRONELLOL ✲ REUNION GERANIUM OIL

RIBOFLAVINE

Products: Added to peanut butter, cereals, bread stuffings, flour, grits, cornmeal, pastas and bread products. Occurs naturally in milk, eggs, and organ meats.

Use: As a color additive, dietary supplement, and nutritional supplement.

Health Effects: FDA states GRAS (generally regarded as safe).

Synonyms: CAS: 83-88-5 ✲ BEFLAVINE ✲ 6,7-DIMETHYL-9-*d*-RIBITYLISOALLOXAZINE ✲ 7,8-DIMETHYL-10-*d*-RIBITYLISOALLOXAZINE ✲ 7,8-DIMETHYL-10-(*d*-RIBO-2,3,4,5-TETRAHYDROXYPENTYL)ISOALLOXAZINE ✲ FLAVAXIN ✲ HYFLAVIN ✲

HYRE ✲ LACTOFLAVIN ✲ LACTOFLAVINE ✲ RIBIPCA ✲ RIBODERM ✲ RIBOFLAVIN ✲ RIBOFLAVINEQUINONE ✲ VITAMIN B2 ✲ VITAMIN G

RICE BRAN WAX

Products: In candy, chewing gum, fruits (fresh), and vegetables (fresh).

Use: As a coating agent, masticatory (chewing) substance, and release agent.

Health Effects: FDA approves use within limitations.

Synonyms: None found.

RICINOLEIC ACID

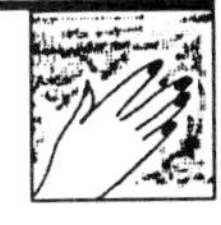

Products: In soaps, dry-cleaning soaps, castor oil, fabric sizing, and in skin softener lotions.

Use: For fabric conditioning, cleaning, and finishing. For skin moisturizing.

Health Effects: A possible carcinogen (cancer-causing agent).

Synonyms: CAS: 141-22-0 ✲ RICINOLIC ACID ✲ RICINIC ACID ✲ CASTOR OIL ACID ✲ 12-HYDROXYOLEIC ACID

ROSIN

Products: In adhesives, varnishes, soaps, inks, rosin bags, and mastics. A product of pine trees.

Use: Various.

Health Effects: Harmless when used for intended purposes.

Synonyms: ✲ GUM ROSIN ✲ WOOD ROSIN ✲ TALL OIL ROSIN

ROTENONE

Products: Insecticide, fish poison, fleas powders, fly sprays, and mothproofing agents.

Use: A multi-purpose pesticide.

Health Effects: Human poison by swallowing. A skin and eye irritant. Acute poisoning causes numbness, nausea, vomiting and tremors.

Synonyms: CAS: 83-79-4 ✲ TUBATOXIN ✲ BARBASCO ✲ DACTINOL ✲ DERIL ✲ DRI-KIL ✲ NOXFISH ✲ CUREX FLEA DUSTER ✲ ROTOCIDE

RUE OIL and HERB

Products: As fruit and spice flavoring in baked goods, candy (soft), frozen

Use: An ingredient that affects taste and smell; a seasoning.

dairy desserts, and mixes.

Health Effects: A possible allergen. FDA states GRAS (generally regarded as safe) when used within limitations.

Synonyms: None found.

S

"Saccharin, the most widely known non-nutritive sweetener, has been around since 1879."

Medical Times/Dr. Allan Bruckheim

SACCHARIN

Products: In bacon, bakery products, beverage mixes, beverages, chewing gum, desserts, fruit juice drinks, jam, relishes, breath fresheners, and chewable vitamin tablets.

Use: A nonnutritive synthetic sweetener.

Health Effects: Slightly toxic by swallowing excessive quantites of pure substance. A human mutagen (changes inherited characteristics). FDA permits use within limitations. The National Academy of Sciences has stated that saccharin is a potential carcinogen (causes cancer) and products containing it must have a warning label.

Synonyms: CAS: 128-44-9 ✲ CRYSTALLOSE ✲ SACCHARINE, SODIUM ✲ SAXIN ✲ SODIUM BENZOSULPHIMIDE ✲ SOLUBLE GLUSIDE ✲ SUCCARIL ✲ SUCRA ✲ SYKOSE

SAFFLOWER OIL

Products: In resins, paints, varnishes, dietetic foods, margarine, and hydrogenated shortening.

Use: As coating agent, emulsifying agent (stabilizes and maintains mixes, formulation aid, and texturizing agent.

Health Effects: A possible skin irritant. Swallowing of large doses can cause vomiting. GRAS (generally regarded as safe).

Synonyms: CAS: 8001-23-8 ✲ SAFFLOWER OIL (UNHYDROGENATED)

SAFROL

Products: In cosmetics, soaps, and perfumes.

Use: As a fragrance (sassafras or camphorwood aroma).

Health Effects: A confirmed carcinogen (causes cancer). Moderately toxic by swallowing. A mutagen (changes inherited characteristics). A skin irritant. FDA prohibits use in food.

Synonyms: CAS: 94-59-7 ❋ 5-ALLYL-1,3-BENZODIOXOLE ❋ ALLYLCATECHOL METHYLENE ETHER ❋ ALLYLDIOXYBENZENE METHYLENE ETHER ❋ 1-ALLYL-3,4-METHYLENEDIOXYBENZENE ❋ 4-ALLYL-1,2-METHYLENEDIOXYBENZENE ❋ m-ALLYLPYROCATECHIN METHYLENE ETHER ❋ 4-ALLYLPYROCATECHOL FORMALDEHYDE ACETAL ❋ ALLYLPYROCATECHOL METHYLENE ETHER ❋ 1,2-METHYLENEDIOXY-4-ALLYLBENZENE ❋ 3,4-METHYLENEDIOXY-ALLYBENZENE ❋ 5-(2-PROPENYL)-1,3-BENZODIOXOLE ❋ RHYUNO OIL ❋ SAFROLE ❋ SAFROLE MF ❋ SHIKIMOLE ❋ SHIKOMOL

SALICYLIC ACID

Products: In aspirin and over-the-counter drugs.

Use: As an analgesic (pain-reliever).

Health Effects: Poison by swallowing large amounts. Effects on the human body by skin contact are ear tinnitus (ringing). A mutagen (changes inherited characteristics). A skin and severe eye irritant.

Synonyms: CAS: 69-72-7 ❋ o-HYDROXYBENZOIC ACID ❋ 2-HYDROXYBENZOIC ACID ❋ KERALYT ❋ ORTHOHYDROXYBENZOIC ACID ❋ RETARDER W ❋ SA

SAPONIN

Products: In soaps, shampoos, detergents, shaving creams, fire extinguishers, detergents, and bath products.

Use: As a foam producer in consumer products; an emulsifier.

Health Effects: Very low toxicity when swallowed.

Synonyms: SAPONARIA ❋ QUILLAJA

SASSAFRAS

Products: In bakery products, beverages (nonalcoholic), confections, gelatin desserts, and puddings. In soaps, toothpastes, perfumes, and powders.

Use: A seasoning and scenting agent. An ingredient that affects taste or smell of product.

Health Effects: A skin irritant. FDA approves use at moderate levels to accomplish the intended effect.

Synonyms: SASSAFRAS ALBIDUM

SAVORY OIL (summer variety)

Products: In salads, sauces, meat, poultry stuffing, and soups.

Use: An aromatic mint used as a seasoning.

Health Effects: Moderately toxic by swallowing excessive amounts of concentrated oil, and possibly by skin contact, of pure substance. A severe skin irritant. FDA states GRAS (generally regarded as safe) when used at moderate levels.

Synonyms: CAS: 8016-68-0 ✲ *SATURIEA HORTENSIS L.*

SEBACIC ACID

Products: In paint products, candles, perfumes, and lubricants.

Use: Various.

Health Effects: Moderately toxic by swallowing.

Synonyms: CAS: 111-20-6 ✲ SEBACYLIC ACID ✲ DECANEDIOIC ACID ✲ 1,8-OCTANEDICARBOXYLIC ACID

SHELLAC

Products: Derived from the (almost continuous) excrement of insects that feed on resiniferous trees in India. Found in varnishes, self polishing waxes, abrasives, hair sprays, sealing wax, cements, sealers, cake glazes, confectioneries, food supplements in tablet form, and gum.

Use: As a coating agent, color fixative, food/candy glaze, and surface-finishing agent for floors and furniture.

Health Effects: Could cause allergic reaction in susceptible individuals.

Synonyms: WHITE SHELLAC ✲ REGULAR BLEACHED SHELLAC ✲ WAX-FREE SHELLAC ✲ REFINED BLEACH SHELLAC ✲ LAC ✲ GARNET LAC ✲ GUM LAC ✲ STICK LAC

SILICA

Products: In bacon (cured), baking powder, beer, coffee whiteners, egg yolk (dried), flour, fruits, gelatin desserts, pudding mixes, salt, soups (powdered), tortilla chips, vanilla powder, and vegetables. In cosmetics, pharmaceuticals, and food.

Use: As an anticaking agent, antifoaming agent, carrier, malt beverage chillproofing agent, color fixative, conditioning agent, defoaming agent, ink (food marking), and waxes to prevent slipping.

Health Effects: Moderately toxic by swallowing excessive quantities of pure chemical. Much less toxic than crystalline forms. Does not cause silicosis. FDA permits use within limitations.

Synonyms: CAS: 112945-52-5 ✲ ACTICEL ✲ AEROSIL ✲ AMORPHOUS SILICA DUST ✲ AQUAFIL ✲ CAB-O-GRIP II ✲ CAB-O-SIL ✲ CAB-O-SPERSE ✲ CATALOID ✲ COLLOIDAL SILICA ✲ COLLOIDAL SILICON DIOXIDE ✲ DICALITE ✲ DRI-DIE INSECTICIDE 67 ✲ FLO-GARD ✲ FOSSIL FLOUR ✲ FUMED SILICA ✲ FUMED SILICON DIOXIDE ✲ HI-SEL ✲ LO-VEL ✲ LUDOX ✲ NALCOAG ✲ NYACOL ✲ SANTOCEL ✲ SILICA AEROGEL ✲ SILICA, AMORPHOUS ✲ SILICIC ANHYDRIDE ✲ SILICON DIOXIDE ✲ SILIKILL ✲ SYNTHETIC AMORPHOUS SILICA ✲ VULKASIL

SODIUM ALUMINOSILICATE

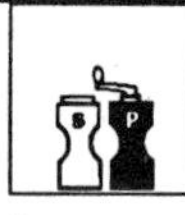

Products: In cake mixes, mixes (dry), nondairy creamers, salt, and sugar (powdered).

Use: An anticaking agent.

Health Effects: FDA states GRAS (generally regarded as safe) within limitations. An irritant to skin, eyes, nose, and throats.

Synonyms: CAS: 1344-00-9 ❊ SODIUM SILICOALUMINATE

SODIUM BENZOATE

Products: In tobacco, creams, carbonated drinks, pickles, toothpastes, fruit juice, preserves, margarine, and oleomargarine.

Use: An antimicrobial (kills germs) agent; preservative for food and pharmaceuticals.

Health Effects: FDA states GRAS (generally regarded as safe) when used within limitations. Moderately toxic by swallowing large amounts of pure chemical. Small doses have little or no effect.

Synonyms: CAS: 532-32-1 ❊ ANTIMOL ❊ BENZOATE of SODA ❊ BENZOATE SODIUM ❊ BENZOESAEURE ❊ BENZOIC ACID, SODIUM SALT ❊ SOBENATE ❊ SODIUM BENZOIC ACID

SODIUM BICARBONATE

Products: In baked goods, beverages (dry mix), fats (rendered), margarine, oleomargarine, pickles (cured), soups, poultry, dairy products, and vegetables. In mouthwashes, bath products, and skin medication products.

Use: An alkali, gastric antacid, cleaning agent, and leavening agent.

Health Effects: Effects on the human body by swallowing excessive amounts of the chemical are respiratory changes, increased urine volume, and sodium level changes. A mutagen (changes inherited characteristics). FDA states GRAS (generally regarded as safe) when used in appropriate amounts. USDA approves use sufficient for purpose.

Synonyms: CAS: 144-55-8 ❊ BAKING SODA ❊ BICARBONATE of SODA ❊ CARBONIC ACID MONOSODIUM SALT ❊ COL-EVAC ❊ JUSONIN ❊ MONOSODIUM CARBONATE ❊ NEUT ❊ SODA MINT ❊ SODIUM ACID CARBONATE ❊ SODIUM HYDROGEN CARBONATE

SODIUM BISULFITE

Products: A preservative.

Use: Prevents discolorization to dried fruit, fresh shrimp, dried, fried, frozen potatoes, and prevents bacterial growth in wine.

Health Effects: Moderately toxic by swallowing excessive amounts of pure substance. A corrosive irritant to skin, eyes, nose and throats. A mutagen (changes inherited characteristics). Causes allergic reaction in susceptible individuals. GRAS (generally regarded as safe).

Synonyms: CAS: 7631-90-5 ❖ BISULFITE de SODIUM (FRENCH) ❖ HYDROGEN SULFITE SODIUM ❖ SODIUM ACID SULFITE ❖ SODIUM BISULFITE ❖ SODIUM BISULFITE (1:1) ❖ SODIUM BISULFITE, solid ❖ SODIUM BISULFITE, solution ❖ SODIUM HYDROGEN SULFITE ❖ SODIUM HYDROGEN SULFITE, solid ❖ SODIUM HYDROGEN SULFITE, solution ❖ SODIUM SULHYDRATE ❖ SULFUROUS ACID, MONOSODIUM SALT

SODIUM CARBOXYMETHYL CELLULOSE

Products: In baked pie fillings, poultry, ice cream, beer, icings, diet foods, and candies.

Use: As a binder (to hold substances together), extender, stabilizer (keeps uniform consistency), and thickener (adds body and thickness to texture).

Health Effects: Mildly toxic by swallowing excessive amounts of pure substance. FDA states GRAS (generally regarded as safe) when used within stated limitations.

Synonyms: CAS: 9004-32-4 ❖ CMC ❖ CARMETHOSE ❖ CELLOFAS ❖ CELLUGEL ❖ CELLULOSE GUM ❖ CELLULOSE SODIUM GLYCOLATE ❖ FINE GUM HES ❖ LUCEL (polysaccharide)

SODIUM CASEINATE

Products: In bread, cereals, cheese (imitation), coffee whiteners, desserts, egg substitutes, loaves (nonspecific), meat (processed), poultry, sausage (imitation), soups, stews, whipped toppings, whipped toppings (vegetable oil), and wine.

Use: A milk protein used as a binder (holds substances together), clarifying agent (removes small particles from liquids), emulsifier (stabilizes and maintains mixes), extender, and stabilizer (keeps a uniform consistency).

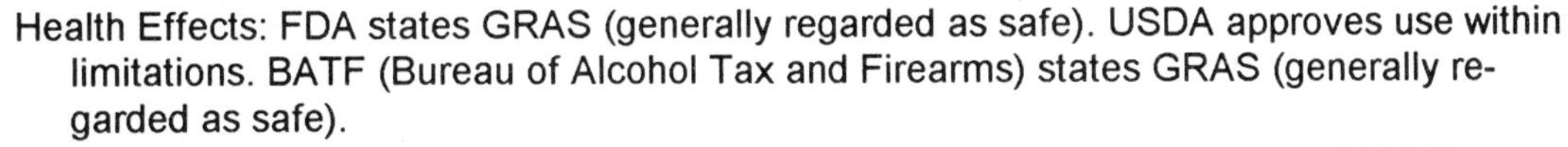

Health Effects: FDA states GRAS (generally regarded as safe). USDA approves use within limitations. BATF (Bureau of Alcohol Tax and Firearms) states GRAS (generally regarded as safe).

Synonyms: CAS: 9005-46-3 ❖ CASEIN and CASEINATE SALTS ❖ CASEIN-SODIUM ❖ CASEIN, SODIUM COMPLEX ❖ CASEINS, SODIUM COMPLEXES ❖ NUTROSE

SODIUM CHLORIDE

Products: In over-the-counter pharmaceuticals, antiseptics, and astringents for skin abrasions. In soaps, bath, and dental products. In baked goods, butter, cheese, nuts (salted), poultry, and

Use: As chilling media, curing agent, dough conditioner, flavoring agent, intensifier, nutrient, and preservative. For ice and snow control.

sausage. In mineral waters and home water softeners.

Health Effects: Moderately toxic by swallowing excessive amounts of chemical. An effect on the human body by swallowing is blood pressure increase. A skin and eye irritant. Swallowing of large amounts of sodium chloride can cause irritation of the stomach. Improper use of salt tablets may produce this effect. USDA states GRAS (generally regarded as safe) when used within limits.

Synonyms: CAS: 7647-14-5 ✲ COMMON SALT ✲ DENDRITIS ✲ EXTRA FINE 200 SALT ✲ EXTRA FINE 325 SALT ✲ HALITE ✲ H.G. BLENDING ✲ NATRIUMCHLORID (GERMAN) ✲ PUREX ✲ ROCK SALT ✲ SALINE ✲ SALT ✲ SEA SALT ✲ STERLING ✲ TABLE SALT ✲ TOP FLAKE ✲ USP SODIUM CHLORIDE ✲ WHITE CRYSTAL

SODIUM CITRATE

Products: In soft drinks, frozen desserts, meat products, detergents, and special cheeses.

Use: As a buffer (in carbonated beverages); improves whipping properties in cream; an emulsifier (in cheese). Prevents separation of solids in evaporated milk.

Health Effects: Harmless when used for intended purposes.

Synonyms: CAS: 68-04-2 ✲ TRISODIUM CITRATE

SODIUM DIACETATE

Products: In baked goods, bread, candy (soft), fats, gravies, meat products, oil, sauces, snack foods, and soups.

Use: As antimicrobial agent (antibacterial), flavoring agent, mold inhibitor, preservative, and sequestrant (binds constituents that affect the final products appearance, flavor or texture).

Health Effects: FDA states GRAS (generally regarded as safe) when used within limits.

Synonyms: CAS:126-96-5 ✲ SODIUM HYDROGEN DIACETATE

SODIUM DICHLOROCYANURATE

Products: In dry bleaches, dishwashing detergents, scouring powder, detergent sanitizers, and pool disinfectants.

Use: As a replacement for calcium hypochlorite and bleaching materials.

Health Effects: Moderately toxic by swallowing concentrated chemical. A skin and severe eye irritant. Swallowing causes emaciation, weakness, lethargy, diarrhea, and weight loss.

Synonyms: CAS: 2244-21-5 ✲ SODIUM SALT OF DICHLORO-S-TRIAZINE-2,4,6-TRIONE

SODIUM DODECYL SULFATE

Products: In carpet shampoos, detergents, and wax products.

Use: In cleaning products and soil emulsifiers.

Health Effects: Moderately toxic by swallowing. A skin irritant. A mild allergen.

Synonyms: CAS: 151-21-3 ❋ AQUAREX METHYL ❋ DODECYL ALCOHOL, HYDROGEN SULFATE, SODIUM SALT ❋ DODECYL SODIUM SULFATE ❋ DODECYLSULFATE SODIUM SALT ❋ DREFT ❋ LAURYL SODIUM SULFATE ❋ SODIUM MONODECYL SULFATE ❋ SULFURIC ACID, MONODECYL ESTER, SODIUM SALT

SODIUM HYDROXIDE

Products: In drain cleaners, oven cleaners, caustic sodas, lye. In black olives, brandy, margarine, meat food products, oleomargarine, poultry products, wine spirit (brandies, sherries, etc.). In hair processing (straightening) products, cuticle removers, and shaving cream products. Emulsifier in liquid cosmetics.

Use: Various.

Health Effects: Moderately toxic by swallowing concentrated amounts. A mutagen (changes inherited characteristics). A corrosive irritant to skin, eyes, nose, and throats. This material, both solid and in solution, has a markedly corrosive action upon all body tissue causing burns and frequently deep ulceration, with ultimate scarring. Mists, vapors, and dusts of this compound cause small burns, and contact with the eyes rapidly causes severe damage to the delicate tissue. Swallowing causes very serious damage to the nose and throats or other tissues with which contact is made. It can cause perforation and scarring. Breathing of the dust or concentrated mist can cause damage to the upper respiratory tract and to lung tissue, depending upon the severity of the exposure. Thus, effects of breathing may vary from mild irritation of the nose and throat to a severe pneumonitis.

Synonyms: CAS: 1310-73-2 ❋ CAUSTIC SODA ❋ CAUSTIC SODA, bead ❋ CAUSTIC SODA, dry ❋ CAUSTIC SODA, flake ❋ CAUSTIC SODA, granular ❋ CAUSTIC SODA, liquid ❋ CAUSTIC SODA, solid ❋ CAUSTIC SODA, solution ❋ LEWIS-RED DEVIL LYE ❋ LYE ❋ SODA LYE ❋ SODIUM HYDRATE ❋ SODIUM HYDROXIDE, bead ❋ SODIUM HYDROXIDE, dry ❋ SODIUM HYDROXIDE, flake ❋ SODIUM HYDROXIDE, granular ❋ SODIUM HYDROXIDE, solid ❋ WHITE CAUSTIC

SODIUM HYPOCHLORITE

Products: Bleach, disinfectant, and drain cleaners.

Use: As swimming pool disinfectant, laundry bleaches, water purification, and germicide.

Health Effects: A human mutagen (changes inherited characteristics). Corrosive and irritating by swallowing and breathing.

Synonyms: CAS: 7681-52-9 ❊ ANTIFORMIN ❊ CHLOROS ❊ CHLOROX ❊ HYCLORITE ❊ SURCHLOR

SODIUM LAURYL SULFATE

Products: In angel food cake, beverage bases, citrus fruit (fresh), egg white (frozen), egg white (liquid), egg white solids, fruit juice drink, marshmallows, poultry, and vegetable oils. In detergents, shampoos, toothpastes, hand lotions, and bubble baths.

Use: An emulsifier (stabilizes and maintains mixes), surfactant, wetting agent, and whipping agent.

Health Effects: Moderately toxic by swallowing excessive amounts of pure substance. A human skin irritant. A mild allergen. A mutagen (changes inherited characteristics).

Synonyms: CAS: 151-21-3 ❊ AQUAREX METHYL ❊ CARSONOL SLS ❊ CONCO SULFATE WA ❊ CYCLORYL 21 ❊ DETERGENT 66 ❊ DODECYL ALCOHOL, HYDROGEN SULFATE, SODIUM SALT ❊ DODECYL SODIUM SULFATE ❊ DODECYL SULFATE, SODIUM SALT ❊ DREFT ❊ DUPONOL ❊ HEXAMOL SLS ❊ IRIUM ❊ LANETTE WAX-S ❊ LAURYL SODIUM SULFATE ❊ LAURYL SULFATE, SODIUM SALT ❊ MAPROFIX 563 ❊ NEUTRAZYME ❊ ORVUS WA PASTE ❊ SIPEX OP ❊ SIPON WD ❊ SLS ❊ SODIUM DODECYL SULFATE ❊ SODIUM MONODODECYL SULFATE ❊ SOLSOL NEEDLES ❊ STERLING WAQ-COSMETIC ❊ SULFOPON WA 1 ❊ SULFOTEX WALA ❊ SULFURIC ACID, MONODODECYL ESTER, SODIUM SALT ❊ ULTRA SULFATE SL-1

SODIUM METAPHOSPHATE

Products: In tooth pastes, detergents, and water softeners. In cheese, dairy products, fish, lima beans, meat food products, milk, peanuts, peas (canned), and poultry food products.

Use: As a sequestrant (binds constituents that affect the final products appearance, flavor or texture), emulsifier (stabilizes and maintains mixes), additive, and for laundering.

Health Effects: FDA states GRAS (generally regarded as safe) when used within limits.

Synonyms: CAS: 10361-03-2 ❊ GRAHAM'S SALT ❊ METAFOS ❊ SODIUM HEXAMETAPHOSPHATE ❊ SODIUM POLYPHOSPHATES, GLASSY ❊ SODIUM TETRAPOLYPHOSPHATE

SODIUM NITRITE

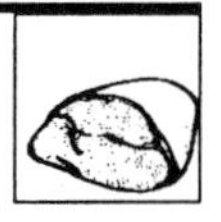

Products: In smoked cured, tunafish, sablefish, salmon, shad, frankfurters, luncheon meats, bacon, corned beef, ham (canned), meat (cured), and poultry.

Use: An antimicrobial (kills germs) agent; color fixative in meat and meat products; a preservative.

Health Effects: Human poison by swallowing large amounts of pure substance. Effects on the human body by swallowing are motor activity changes, coma, decreased blood pressure with possible pulse rate increase without fall in blood pressure, arteriolar or venous dilation, nausea or vomiting, and blood chemistry changes. Nitrite can lead to the formation of small amounts of cancer-causing chemicals (nitrosamines), particularly in fried bacon. Nitrite is tolerated in foods because it prevents growth of bacteria that cause botulism (poisoning). A possible carcinogen (causes cancer). A human mutagen (changes inherited characteristics). FDA permits use within stated limitations.

Synonyms: CAS: 7632-00-0 ✲ ANTI-RUST ✲ DIAZOTIZING SALTS ✲ ERINITRIT ✲ FILMERINE ✲ NITROUS ACID, SODIUM SALT

SODIUM PALMITATE

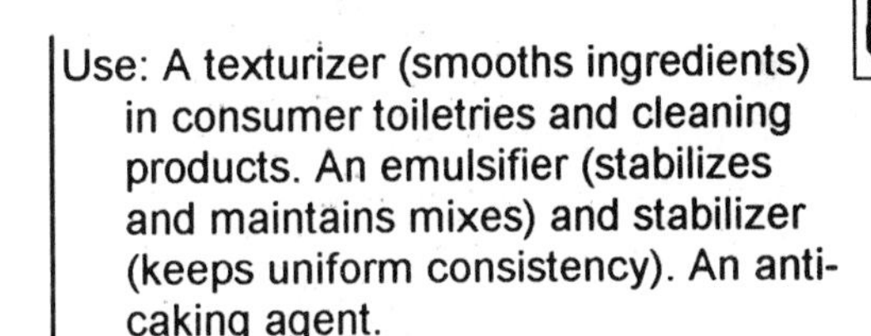

Products: In shaving creams, laundry soaps, toilet soaps, detergents, cosmetics, and inks.

Use: A texturizer (smooths ingredients) in consumer toiletries and cleaning products. An emulsifier (stabilizes and maintains mixes) and stabilizer (keeps uniform consistency). An anti-caking agent.

Health Effects: Harmless when used for intended purposes. FDA states GRAS (generally regarded as safe).

Synonyms: SODIUM SALT OF PALMITIC ACID

SODIUM PERBORATE

Products: For dentures or partial dental plates.

Use: To sanitize, deodorize, and clean. As a bleaching agent in special detergents.

Health Effects: Toxic by swallowing. Strong oxidizing agent (gives off oxygen).

Synonyms: CAS: 7632-04-4 ✲ SODIUM PERBORATE ANHYDROUS ✲ SODIUM PERBORATE MONOHYDRATE

SODIUM PERCARBONATE

Products: For dentures or partial dental plates.

Use: To sanitize, deodorize, and clean. A mild antiseptic. A domestic bleaching agent.

Health Effects: Toxic by swallowing. Strong oxidizing agent (gives off oxygen).

Synonyms: CAS: 4452-58-8 ✲ SODIUM CARBONATE PEROXOHYDRATE

SODIUM PEROXIDE

Products: In various bleaches, deodorants, antiseptics, water purification, dyes, and germicidal soaps. For

Use: Various.

purifying air in sick rooms.

Health Effects: A severe irritant to skin, eyes, nose, and throats. Dangerous fire and explosion risk.

Synonyms: CAS: 1313-60-6 ❖ DISODIUM DIOXIDE ❖ DISODIUM PEROXIDE ❖ SODIUM DIOXIDE ❖ SOLOZONE ❖ SODIUM OXIDE

SODIUM PROPIONATE

Products: In baked goods, beverages (nonalcoholic), candy (soft), cheese, confections, dough (fresh pie), fillings, frostings, gelatins, jams, jellies, meat products, pizza crust, and puddings.

Use: An antimicrobial (kills germs) agent, additive, flavoring agent, mold/mildew inhibitor, and preservative.

Health Effects: An allergen. FDA and USDA permits use within limitations.

Synonyms: CAS: 137-40-6 ❖ PROPANOIC ACID, SODIUM SALT

SODIUM SACCHARIN

Products: Sweetener (nonnutritive).

Use: In bacon, bakery products (nonstandardized), beverage mixes, beverages, chewing gum, desserts, fruit juice drinks, jam, relishes, and vitamin tablets (chewable).

Health Effects: FDA permits use within limitations. Moderately toxic by swallowing concentrated excessive amounts. A human mutagen (changes inherited characteristics).

Synonyms: CAS: 128-44-9 ❖ SWEETA ❖ CRISTALLOSE ❖ CRYSTALLOSE ❖ DAGUTAN ❖ KRISTALLOSE ❖ MADHURIN ❖ ODA ❖ SACCHARIN ❖ SACCHARIN SOLUBLE ❖ SACCHARIN, SODIUM ❖ SACCHARIN, SODIUM SALT ❖ SACCHARINE SOLUBLE ❖ SACCHARINNATRIUM ❖ SACCHAROIDUM NATRICUM ❖ SAXIN ❖ SODIUM-1,2 BENZISOTHIAZOLIN-3-ONE-1,1-DIOXIDE ❖ SODIUM o-BENZOSULFIMIDE ❖ SODIUM BENZOSULPHIMIDE ❖ SODIUM-o-BENZOSULPHIMIDE ❖ SODIUM-2-BENZOSULPHIMIDE ❖ SODIUM SACCHARIDE ❖ SODIUM SACCHARINATE ❖ SODIUM SACCHARINE ❖ SOLUBLE GLUSIDE ❖ SOLUBLE SACCHARIN ❖ SUCCARIL ❖ SUCRA ❖ o-SULFONBENZOIC ACID IMIDE SODIUM SALT ❖ SULPHOBENZOIC IMIDE, SODIUM SALT ❖ SYKOSE ❖ WILLOSETTEN

SODIUM SILICATE

Products: For eggs, cosmetics, soaps, depilatories, protective creams, topical antiseptics, gels, and adhesives.

Use: As a preservative, anticaking agent, and detergent.

Health Effects: Poison by swallowing excessive amounts of concentrated substance. A caustic material which is a severe eye, skin, nose, and throat irritant. Swallowing causes gastrointestinal (stomach and intestinal) tract upset. FDA and USDA approves use at moderate levels.

Synonyms: CAS: 6834-92-0 ✤ B-W ✤ CRYSTAMET ✤ DISODIUM METASILICATE ✤ DISODIUM MONOSILICATE ✤ METSO 20 ✤ METSO BEADS 2048 ✤ METSO BEADS, DRYMET ✤ METSO PENTABEAD 20 ✤ ORTHOSIL ✤ SODIUM METASILICATE ✤ SODIUM METASILICATE, anhydrous ✤ WATER GLASS

SORBIC ACID

Products: In baked goods, beverages (carbonated), beverages (still), bread, cake batters, cake fillings, cake topping, cheese, cottage cheese (creamed), fish (smoked or salted), fruit juices (fresh), fruits (dried), margarine, oleomargarine, pickled goods, pie crusts, pie fillings, salad dressings, salads (fresh), sausage (dry), sea food cocktail, syrups (chocolate dairy), and wine. Not permitted in cooked sausage and meat salads.

Use: Preservative and food additive. Prevents mold growth.

Health Effects: Mildly toxic by swallowing excessive amounts of pure substance. A severe human skin irritant. A mutagen (changes inherited characteristics). FDA and USDA approves use within limitations.

Synonyms: CAS: 110-44-1 ✤ (2-BUTENYLIDENE)ACETIC ACID ✤ CROTYLIDENE ACETIC ACID ✤ HEXADIENIC ACID ✤ HEXADIENOIC ACID ✤ 2,4-HEXADIENOIC ACID ✤ trans-trans-2,4-HEXADIENOIC ACID ✤ 1,3-PENTADIENE-1-CARBOXYLIC ACID ✤ 2-PROPENYLACRYLIC ACID ✤ SORBISTAT

SORBITAN MONOSTEARATE

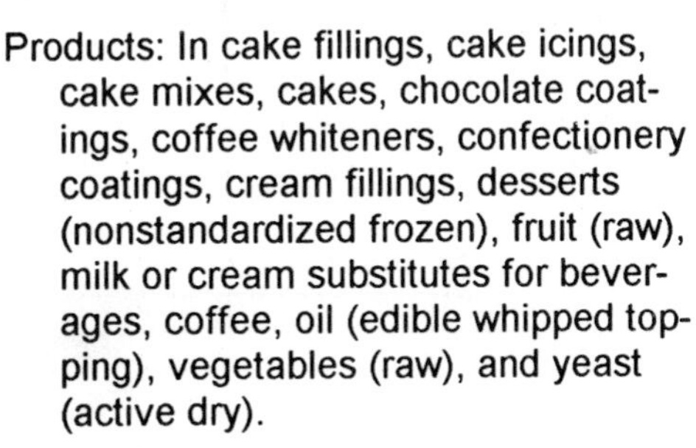

Products: In cake fillings, cake icings, cake mixes, cakes, chocolate coatings, coffee whiteners, confectionery coatings, cream fillings, desserts (nonstandardized frozen), fruit (raw), milk or cream substitutes for beverages, coffee, oil (edible whipped topping), vegetables (raw), and yeast (active dry).

Use: As a defoaming agent, emulsifier (stabilizes and maintains mixes), rehydration aid (replaces moisture), and stabilizer (helps keep a uniform consistency).

Health Effects: Mildly toxic by swallowing large quantities of pure chemical. A skin irritant. FDA approves use within stated limits.

Synonyms: CAS: 1338-41-6 ✤ ANHYDRO-*d*-GLUCITOL MONOOCTADECANOATE ✤ ANHYDROSORBITOL STEARATE ✤ ARLACEL ✤ ARMOTAN ✤ CRILL ✤ DREWSORB ✤ HODAG ✤ IONET ✤ LIPOSORB S-20 ✤

MONTANE ❊ NEWCOL ❊ NIKKOL ❊ SORBITAN C ❊ SORBITAN MONOOCTADECANOATE ❊ SORBITAN STEARATE

SORBITOL

Products: In cosmetic creams, lotions, toothpaste, tobacco, gum, and candy. In baked goods, baking mixes, beverages (low calorie), candy (hard), candy (soft), chewing gum, chocolate, coconut (shredded), cough drops, frankfurter (labeled), frozen dairy desserts, jams (commercial nonstandardized), jellies (commercial nonstandardized), knockwurst, sausage (cooked), and wieners.

Use: An anticaking agent, sweetener, curing agent, moisture-conditioning, drying agent, emulsifier (stabilizes and maintains mixes), firming agent, flavoring agent, formulation aid, free-flow agent, humectant (moisturizer), lubricant, sweetener (nutritive), pickling agent, release agent, sequestrant (affects the final products appearance, flavor or texture), stabilizer (helps to keep uniform consistency), surface-finishing agent, texturizing (smoothing) agent, and thickening agent.

Health Effects: Pure chemical is mildly toxic by swallowing excessive amounts. Effect on the human body by swallowing is diarrhea. FDA states GRAS (generally regarded as safe) when used within limitations. Diabetics find it useful as a result of slow absorption because blood sugar does not increase rapidly.

Synonyms: CAS: 50-70-4 ❊ CHOLAXINE ❊ DIAKARMON ❊ GLUCITOL ❊ *d*-GLUCITOL ❊ GULITOL ❊ *l*-GULITOL ❊ KARION ❊ NIVITIN ❊ SIONIT ❊ SIONON ❊ SORBICOLAN ❊ SORBITE ❊ *d*-SORBITOL ❊ SORBO ❊ SORBOL ❊ SORBOSTYL ❊ SORVILANDE

SOYBEAN OIL

Products: In candy, cooking oil, salad oil, margarine, shortenings, soap, paints, varnishes, and ink.

Use: As a coating agent, emulsifying agent (stabilizes and maintains mixes), formulation aid, and texturizer (used to improve smoothness).

Health Effects: GRAS (generally regarded as safe) when used for intended purposes.

Synonyms: SOYA BEAN OIL ❊ CHINESE BEAN OIL ❊ SOY OIL

SPEARMINT

Products: In ice cream, candies, bakery products, spices, gums, condiments, jellies, beverages, perfumes, cosmetics, and toothpastes.

Use: A flavoring and odorant. An ingredient that affects the taste or smell of products.

Health Effects: Mildly toxic by swallowing excessive amounts of pure substance. A mutagen (changes inherited characteristics). A skin irritant and an allergen. FDA states GRAS (generally regarded as safe).

Synonyms: CAS: 8008-79-5 ✲ OIL of SPEARMINT ✲ COMMON SPEARMINT ✲ SCOTCH SPEARMINT ✲ GARDEN MINT ✲ GREEN MINT

SPIKE LAVENDER OIL

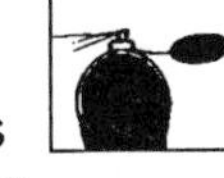

Products: In beverages, ice creams, candies, bakery products; also in perfumes, shaving lotions, soaps, and colognes.

Use: A seasoning or scenting agent. As an ingredient that affects the taste or smell of consumer products.

Health Effects: Concentrated substances are moderately toxic by swallowing large amounts. A skin irritant. FDA states GRAS (generally regarded as safe) when used at moderate levels to accomplish the intended effect.

Synonyms: CAS: 84837-04-7 ✲ LAVENDER OIL, SPIKE ✲ OIL of SPIKE LAVENDER

SQUALANE

Products: In pharmaceuticals, over-the-counter drug products, cosmetics, and perfumes.

Use: A synthetic lubricating oil and vehicle base for consumer products. Prevents drying and congealing of products.

Health Effects: Could cause allergic reaction in susceptible individuals.

Synonyms: CAS: 111-01-3 ✲ PERHYDROSQUALENE ✲ SPINACANE ✲ DODECAHYDROSQUALANE ✲ HEXAMETHYLTETRACOSANE

SQUALENE

Products: A natural raw material found in human sebum (skin oil) and in shark liver oil.

Use: As a lubricating oil; in perfume and hair products.

Health Effects: Could cause allergic reaction to susceptible individuals.

Synonyms: CAS: 7683-64-9 ✲ SPINACENE

STANNIC CHLORIDE

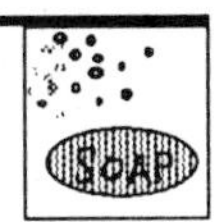

Products: In metallic hair dyes, soaps, and perfumes.

Use: For bacteria and fungi control; for perfume stabilizer.

Health Effects: Moderately toxic by breathing. A corrosive irritant to skin, eyes, and nose and throats.

Synonyms: CAS: 7646-78-8 ❖ TIN CHLORIDE ❖ TIN PERCHLORIDE ❖ TIN TETRACHLORIDE ❖ LIBAVIUS FUMING SPIRIT

STANNIC OXIDE

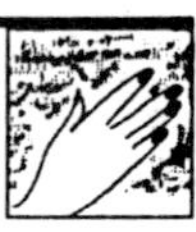

Products: In fingernail polish, glass polish, putty, perfumes, and cosmetics.

Use: A mineral used in polishes and in ceramics.

Health Effects: Harmless when used for intended purposes.

Synonyms: CAS: 18282-10-5 ❖ CASSITERITE ❖ STANNIC ANHYDRIDE ❖ TIN PEROXIDE ❖ TIN DIOXIDE ❖ STANNIC ACID

STANNOUS FLUORIDE

Products: In toothpaste.

Use: To prevent tooth decay.

Health Effects: Poison by swallowing large quantities. In excessive amounts it is a possible carcinogen (causes cancer) and mutagen (changes inherited characteristics).

Synonyms: CAS: 7783-47-3 ❖ TIN FLOURIDE ❖ FLUORISTAN ❖ STANNOUS FLOURIDE ❖ TIN BIFLUORIDE ❖ TIN DIFLUORIDE

STAPLE FIBER ACETATE

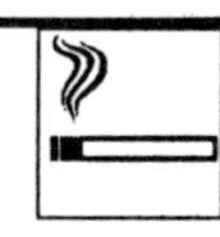

Products: In cigarette filter tips, tobacco smoke filters, and tobacco smoke filter tip rods.

Use: To filter cigarette smoke.

Health Effects: A component of cigarettes. Smoking is a recognized cause of cancer, lung disease, emphysema, and heart problems.

Synonyms: COTTON FIBER

STARCH

Products: In adhesives (gummed paper, tapes, cartons, bags), food products (gravies, custards, confectionery), filler in baking powders (cornstarch), laundry starch, over-the-counter pharmaceuticals, dentifrices, hair colorings, facial rouge, cosmetics, baby powders, dusting powder, and face powder.

Use: As a gelling agent, sizing agent, fabric stiffener, thickener, anticaking agent, and explosives (nitrostarch).

Health Effects: Could cause an allergic reaction. Conversely, it is sometimes used to treat skin irritation.

Synonyms: CAS: 9005-84-9 ❋ CARBOHYDRATE POLYMER

STARCH PHOSPHATE

Products: In frozen foods, adhesives, drugs, and cosmetics.

Use: Soluble in cold water, unlike regular starch. As a thickener.

Health Effects: Harmless when used for intended purposes.

Synonyms: ARABIC GUM SUBSTITUTE ❋ LOCUST BEAN GUM SUBSTITUTE ❋ CARBOXYMETHYL CELLULOSE SUBSTITUTE

STEARIC ACID

Products: In bar soaps, deodorants, cosmetics, facial makeup, hand lotions, ointments, shaving creams, shoe polish, automobile metal polish, skin moisturizers, and suppositories. Also in beverages, bakery products, gum, and candies.

Use: As a defoaming agent, softener, and flavoring agent. For moisturizing, lubricating, and producing pearlized effect in hand creams and soaps.

Health Effects: A human skin irritant. Can cause allergic reaction in susceptible individuals. FDA states GRAS (generally regarded as safe) when used within specifications.

Synonyms: CAS: 57-11-4 ❋ CENTURY ❋ DAR-CHEM ❋ EMERSOL ❋ GLYCON ❋ GROCO ❋ 1-HEPTADECANECARBOXYLIC ACID ❋ HYDROFOL ACID ❋ HY-PHI ❋ HYSTRENE ❋ INDUSTRENE ❋ KAM ❋ NEO-FAT ❋ OCTADECANOIC ACID ❋ PEARL STEARIC ❋ STEAREX BEADS ❋ STEAROPHANIC ACID ❋ TEGOSTEARIC

STEROIDS, ANABOLIC

Products: In drugs and medications.

Use: To promote growth and repair body tissue.

Health Effects: A possible carcinogenic (causes cancer). May be teratogenic (abnormal fetal development) and causes reproductive (infertility, or sterility, or birth defects) effects. Anabolic steroids are synthetic derivatives of testosterone (male hormone). They have beneficial medical benefits when used appropriately for senility, debilitating illness, and for certain convalescents. They are much abused by atheletes and can cause serious harmful effects.

Synonyms: VARIOUS

STRYCHNINE

Products: A rat and mouse poison.

Use: A pesticide.

Health Effects: Lethal human poison. Effects on the human body by swallowing are muscular twitching, convulsions, spasms, and death. On EPA's Extremely Hazardous Substances List.

Synonyms: CAS: 57-24-9 ❖ CERTOX ❖ KWIK-KIL ❖ RO-DEX ❖ SANASEED ❖ STRYCHNOS ❖ MOUSE-TOX ❖ MOUSE-RID

SUCCINIC ACID

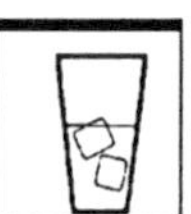

Products: In beverages, condiments, meat products, relishes, and sausage (hot). Also in mouthwash, perfume, dye, and lacquers.

Use: As a flavor enhancer; miscellaneous and general-purpose food chemical; and neutralizing agent (a buffer that eliminates both acidity and alkalinity keeping the product neutral).

Health Effects: A severe eye irritant. FDA states GRAS (generally regarded as safe) when used within limitations.

Synonyms: CAS: 110-15-6 ❖ AMBER ACID ❖ BERNSTEINSAURE (GERMAN) ❖ BUTANEDIOIC ACID ❖ 1,2-ETHANEDICARBOXYLIC ACID ❖ ETHYLENESUCCINIC ACID

SUCROSE

Products: Table sugar available as granulated, brown, and powder sugar.

Use: Sweetener in desserts, beverages, cakes, ice creams, icings, cereals, and baked goods.

Health Effects: Mildly toxic by swallowing excessive quantities of pure substance. GRAS (generally regarded as safe) when used at moderate levels. Provides no vitamins, minerals, or protein. Americans consume over 60 pounds a year.

Synonyms: CAS: 57-50-1 ❖ BEET SUGAR ❖ CANE SUGAR ❖ CONFECTIONER'S SUGAR ❖ α-d-GLUCOPYRANOSYL β-d-FRUCTOFURANOSIDE ❖ (α-d-GLUCOSIDO)-β-d-FRUCTOFURANOSIDE ❖ GRANULATED SUGAR ❖ ROCK CANDY ❖ SACCHAROSE ❖ SACCHARUM ❖ SUGAR

SUCROSE FATTY ACID ESTERS

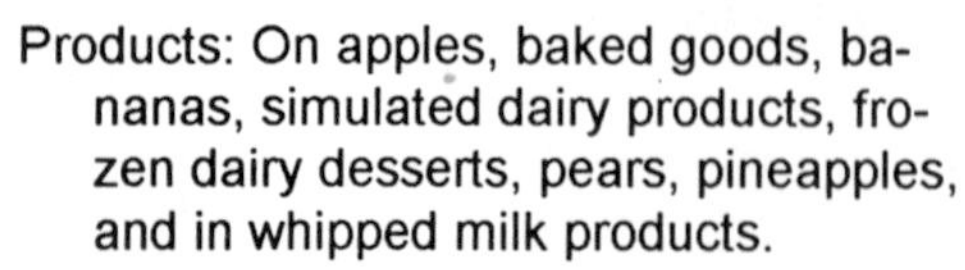

Products: On apples, baked goods, bananas, simulated dairy products, frozen dairy desserts, pears, pineapples, and in whipped milk products.

Use: As an emulsifier (stabilizes and maintains mixes), coating (protective), and as a texturizing agent.

Health Effects: Harmless when used for intended purposes. FDA permits use in moderate levels.

Synonyms: None found.

SULFAMIC ACID

Products: In swimming pool chemicals (stabilizer); flameproofing of fabrics and wood; as a weed killer; and food packaging

Use: Various.

Health Effects: Moderately toxic by swallowing. A human skin irritant. A corrosive irritant to skin, eyes, nose and throats. A substance which migrates to food from packaging materials. FDA states GRAS (generally regarded as safe) when used appropriately.

Synonyms: CAS: 5329-14-6 ❋ AMIDOSULFONIC ACID ❋ AMIDOSULFURIC ACID ❋ AMINOSULFONIC ACID ❋ SULFAMIDIC ACID ❋ SULPHAMIC ACID

SULFITES

Products: In various food and beverage products.

Use: As a preservative and antioxidant (slows down the spoiling of foods, prevents racidity).

Health Effects: Banned from restaurant and supermarket use on salads and vegetables as a result of reactions of individuals. These included unconsciousness, anaphylactic schock, nausea, diarrhea, and asthma attacks which resulted in deaths in some incidents. It is still used in wines and packaged foods, when appropriately labeled.

Synonyms: SULFUR DIOXIDE ❋ SODIUM BISULFITE ❋ SODIUM METABISULFITE ❋ POTASSIUM METABISULFITE

SULFORAPHANE

Products: Primarily in cruciferous vegetables, broccoli, Brussels sprouts, cabbage, and cauliflower.

Use: The chemical triggers enzymes known to neutralize carcinogens in cells.

Health Effects: Very beneficial as an anticancer ingredient.

Synonyms: None known

SULFURIC ACID

Products: Battery acid (EXTREME CAUTION SHOULD BE EXERCISED WHEN CHARGING BATTERIES OR JUMP STARTING VEHICLES).

Use: The electrolyte.

Health Effects: Human poison by unspecified route. Moderately toxic by swallowing. A severe eye irritant. Extremely irritating, corrosive, and toxic to tissue, resulting in destruction of tissue causing severe burns. Repeated contact with dilute solutions can cause a dermatitis, and repeated or prolonged breathing can cause inflammation of the upper respiratory tract leading to chronic bronchitis. Exposure may cause a chemical pneumonia.

Synonyms: CAS: 7664-93-9 ✲ VITRIOL BROWN OIL ✲ OIL OF VITRIOL ✲ HYDROOT ✲ DIPPING ACID ✲ NORDHAUSEN ACID ✲ MATTING ACID ✲ SULPHURIC ACID

SULFUROUS ACID

Products: A fruit, nut, food, and wine preservative. In dental bleaching agents.

Use: As an antiseptic, preservative, and bleach.

Health Effects: Concentrated amounts are toxic by swallowing and breathing.

Synonyms: CAS: 7782-99-2 ✲ 6% SULFUR DIOXIDE SOLUTION

SUNFLOWER OIL

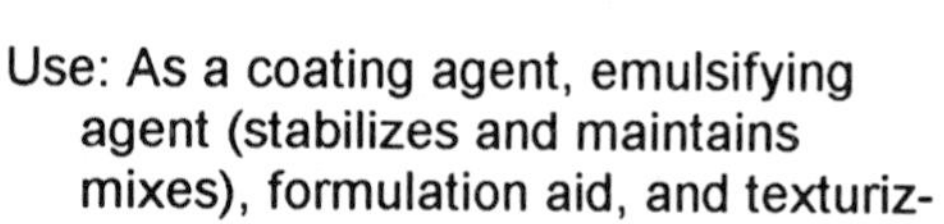

Products: In margarine, shortening, soaps, and dietary supplements.

Use: As a coating agent, emulsifying agent (stabilizes and maintains mixes), formulation aid, and texturizing agent.

Health Effects: Harmless when used for intended purposes.

Synonyms: None found.

SUNSCREENS

Products: In lotions, oils, gels, creams, mousse, sticks, and roll-ons.

Use: To protect the skin from sun exposure and damage. Use a high SPF (sun protection factor) numbered product.

Health Effects: No sunscreen can completely block out the sun's damaging rays. Therefore, it is necessary to limit exposure and reapply after swimming and exercise. The sun's two ultraviolet rays are UVA and UVB. UVA primarily causes premature aging and wrinkling. UVB primarily causes sunburn and skin cancer. Although both can cause aging and cancer.

Synonyms: OXYBENZONE ✲ DIOXYBENZONE ✲ EUSOLAX 6300 ✲ EUSOLAX 8020 ✲ PARSOL1789 ✲ SULISOBENZONE ✲ RED PETROLATUM ✲ TITANIUM DIOXIDE

SYMCLOSENE

Products: In household cleansers, deodorizers, and cleaning products.

Use: As a disinfectant, deodorizer, and chlorinating agent.

Health Effects: Moderately toxic by swallowing. Effects on the human body by swallowing are ulceration or bleeding from the stomach. A severe eye and skin irritant.

Synonyms: CAS: 87-90-1 ✲ TRICHLORINATED ISOCYANURIC ACID ✲ ISOCYANURIC CHLORIDE ✲ TRICHLOROCYANURIC ACID

SYNTHETIC PARAFFIN

Products: On grapefruit, lemons, limes, muskmelons, oranges, sweet potatoes, and tangerines.

Use: As a coating (protective).

Health Effects: FDA approves use at moderate levels to accomplish the intended effect.

Synonyms: SUCCINIC DERIVATIVES

T

"About 1,800,000 poisonings were reported to the American Association of Poison Control Centers last year. Most were accidental poisonings of children under the age of five."
American Association of Poison Control Centers

TAGETES

Products: A yellow coloring derived from flower petals of the Aztec marigold.

Use: Additive to chicken feed to increase yellow color of the skin and eggs of poultry.

Health Effects: Harmless when used for intended purposes although some sensitive individuals could show allergic reactions.

Synonyms: *Tagetes Erecta L.*

TALC

Products: In baby, bath, foot, and face powders. In pharmaceuticals, soaps, cosmetics, eye shadow, rouges, facial makeup, and creams. In paints, putty, lubricants, slate pencils, and crayons.

Use: As an anticaking agent, coating agent, lubricant, release agent, surface-finishing agent, and texturizing agent.

Health Effects: Could be an allergen and a skin irritant. The prolonged or repeated breathing can produce a form of pulmonary fibrosis (talc pneumoconiosis) which may be due to asbestos in powder. A possible carcinogen. A study of women who used talcum powder on sanitary napkins indicated an increased risk of ovarian cancer. The study indicates that talc entered the reproductive tract in this manner.

Synonyms: CAS: 14807-96-6 ❖ AGALITE ❖ AGI TALC ❖ ALPINE TALC USP ❖ ASBESTINE ❖ DESERTALC ❖ FIBRENE ❖ LO MICRON ❖ METRO TALC ❖ MISTRON FROST ❖ MISTRON STAR ❖ MISTRON SUPER FROST ❖ MISTRON VAPOR ❖ PURTALC USP ❖ SIERRA C-400 ❖ SNOWGOOSE ❖ SOAPSTONE ❖ STEAWHITE ❖ STEATITE ❖ SUPREME DENSE ❖ TALCUM

TALL OIL

Products: In drying oil, soaps, greases, and paints.

Use: As a lubricant, paint vehicle, and emulsifier.

Health Effects: A mild allergen. Derived from pine wood. GRAS (generally regarded as safe). An indirect additive from packages. A substance which migrates to food from packaging materials.

Synonyms: CAS: 8002-26-4 ✲ LIQUID ROSIN ✲ TALLOL

TALLOW

Products: In soap stock, leather dressing, candles, greases, and in animal feeds.

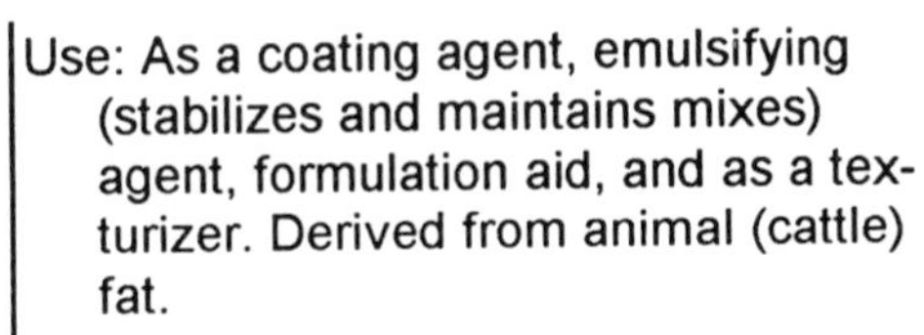

Use: As a coating agent, emulsifying (stabilizes and maintains mixes) agent, formulation aid, and as a texturizer. Derived from animal (cattle) fat.

Health Effects: GRAS (generally regarded as safe) when used for intended purposes.

Synonyms: BLEACHED-DEODORIZED TALLOW

TANGERINE OIL

Products: In desserts, soft drinks, ice cream, furniture polish; an odorant in perfume and cosmetics.

Use: An ingredient that affects the taste or aroma of foods, cosmetics, and cleaning products.

Health Effects: A skin irritant. GRAS (generally regarded as safe) when used at a level not in excess of the amount necessary to accomplish the desired results.

Synonyms: CAS: 8008-31-9 ✲ TANGERINE OIL, COLDPRESSED (FCC) ✲ TANGERINE OIL, EXPRESSESED (FCC) ✲ CITRUS PEEL OIL

TANNIC ACID

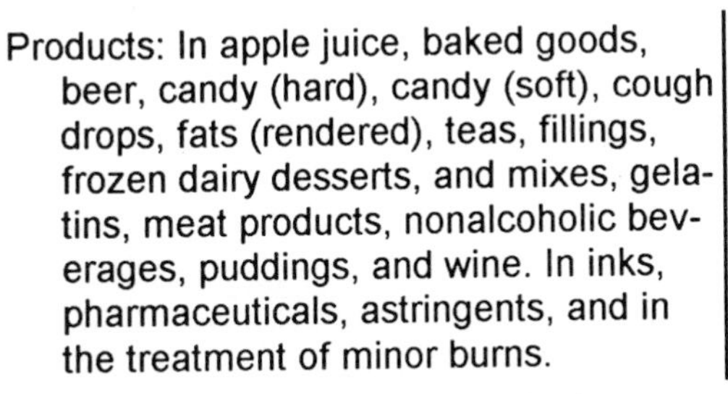

Products: In apple juice, baked goods, beer, candy (hard), candy (soft), cough drops, fats (rendered), teas, fillings, frozen dairy desserts, and mixes, gelatins, meat products, nonalcoholic beverages, puddings, and wine. In inks, pharmaceuticals, astringents, and in the treatment of minor burns.

Use: Derived from tree barks, nutgalls, and plant parts. Used as a clarifying agent in beer and wine; a flavor enhancer and agent.

Health Effects: Poison by swallowing excessive amounts of pure acid. FDA states GRAS (generally regarded as safe) when used within limitations.

Synonyms: CAS: 1401-55-4 ✲ d'ACIDE TANNIQUE (FRENCH) ✲ GALLOTANNIC ACID ✲ GALLOTANNIN ✲ GLYCERITE ✲ TANNIN

TARRAGON OIL

Products: A seasoning oil for liqueurs, salad dressings, soups, and sauces. In perfume.

Use: An herb that affects the taste or smell of food or cologne.

Health Effects: Moderately toxic by swallowing large concentrated amounts. A skin irritant. FDA states GRAS (generally regarded as safe) when used within reasonable limits.

Synonyms: CAS: 8016-88-4 ✣ ESTRAGON OIL

TARTARIC ACID

Products: In baking powder, beverages (grape and lime flavored), jellies (grape flavored), poultry, and wine.

Use: An acid, firming agent, flavor enhancer, flavoring agent, humectant (keeps product from drying out), sequestrant (affects the products appearance, flavor or texture).

Health Effects: Pure chemical is mildly toxic by swallowing excessive quantities. FDA states GRAS (generally regarded as safe) when used within stated limitations.

Synonyms: CAS: 87-69-4 ✣ 2,3-DIHYDROSUCCINIC ACID ✣ 2,3-DIHYDROXYBUTANEDIOC ACID

TCSA

Products: In surgical soaps, laundry soaps, rinses, deodorants, shampoos, and polishes.

Use: As a bacteriostat (kills germs) to prevent bacterial growth in cleaning products.

Health Effects: Poison by swallowing. A skin irritant.

Synonyms: 3,3′,4′,5-TETRACHLOROSALICYLANILDE ✣ 3,5-DICHLORO-n-(3,4-DICHLOROPHENYL)-2-HYDROBENZAMIDE ✣ IRGASAN BS2000

TEA

Products: Hot or cold brewed drink.

Use: As a refreshment or stimulant beverage.

Health Effects: Polyphenol substances in tea are believed to inhibit the action of carcinogens (causes cancer) in food and tobacco. May protect against heart disease and breast cancer. Other elements of tea may prevent tooth decay. Tannins in tea help fight bacteria and viruses

Synonyms: GREEN TEA ✣ OOLONG TEA ✣ BLACK TEA ✣ HERBAL TEAS ✣ GINSENG TEA

TEA LAURYL SULFATE

Products: In detergents, cosmetics, shampoos, and pharmaceuticals.

Use: As a foaming, wetting, and dispersing agent.

Health Effects: A skin and eye irritant.

Synonyms: CAS: 139-96-8 ✲ TRIETHANOLAMINE LAURYL SULFATE ✲ SULFURIC ACID, MONODECYL ESTER ✲ SULFURIC ACID, DODECYL ESTER, TRIETHANOLAMINE SALT

TERPINEOL

Products: In perfumes, soaps, antiseptics, disinfectants, and flavoring agents.

Use: An ingredient to affect the taste or smell of products.

Health Effects: Mildly toxic by swallowing. A skin irritant.

Synonyms: CAS: 8006-39-1 ✲ p-MENTH-1-EN-8-OL ✲ MIXTURE OF p-METHENOLS ✲ α-TERPINEOLS

THIOLACTIC ACID

Products: In depilatories (hair removers), permanent wave solutions, and hair straightening products.

Use: In hair curling, removing, and processing products.

Health Effects: Poisoning by swallowing. Mildly toxic by breathing. Can cause severe allergic reaction and skin irritation.

Synonyms: CAS: 79-42-5 ✲ 2-MERCAPTOPROPIONIC ACID ✲ 2-MERCAPTOPROPANOIC ACID ✲ 2-THIOLPROPIONIC ACID ✲ α-MERCAPTOPROPANOIC ACID

THIOUREA

Products: In photocopy paper, photography, dyes, drugs, and hair preparations.

Use: A fixer; a mold and mildew preventative.

Health Effects: A probable carcinogen and a poison. A human mutagen (changes inherited characteristics). Effects on the human body by swallowing are hemorrhage, blood cell changes, and possible depression of bone marrow with anemia. May cause allergic skin eruptions.

Synonyms: CAS: 62-56-6 ✲ THIOCARBAMIDE

THYMOL

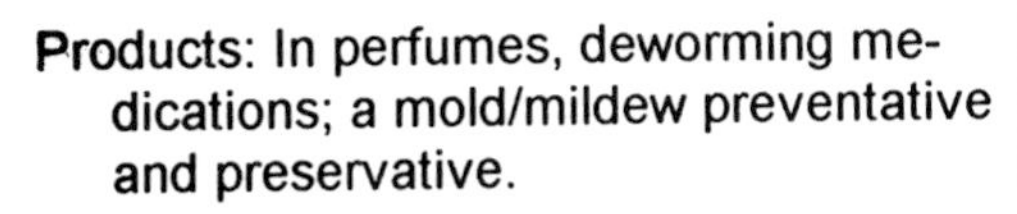

Products: In perfumes, deworming medications; a mold/mildew preventative and preservative.

Use: An antibacterial and antifungal drug.

Health Effects: Poison by swallowing inappropriate amounts. Could cause allergic reaction. An FDA over-the-counter drug.

Synonyms: CAS: 89-83-8 ✲ ISOPROPYL-m-CRESOL ✲ THYME CAMPHOR ✲ THYMIC ACID ✲ 3-p-CYMENOL

TITANIUM DIOXIDE

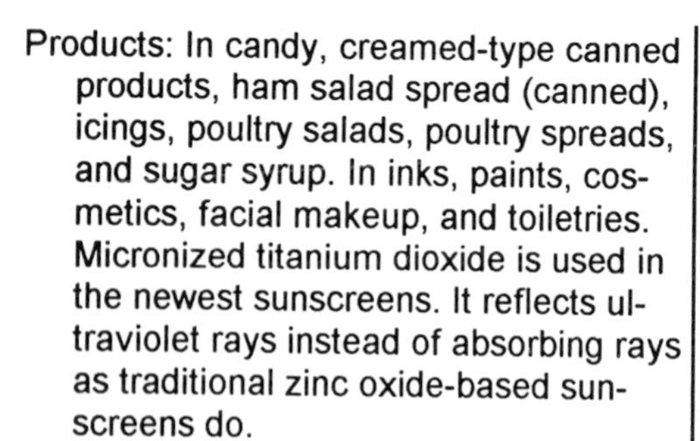

Products: In candy, creamed-type canned products, ham salad spread (canned), icings, poultry salads, poultry spreads, and sugar syrup. In inks, paints, cosmetics, facial makeup, and toiletries. Micronized titanium dioxide is used in the newest sunscreens. It reflects ultraviolet rays instead of absorbing rays as traditional zinc oxide-based sunscreens do.

Use: As a color additive, a filler, sunscreen, and pigment.

Health Effects: A possible skin irritant. FDA and USDA approve use within limitations.

Synonyms: CAS: 13463-67-7 ✲ A-FIL CREAM ✲ ATLAS WHITE TITANIUM DIOXIDE ✲ AUSTIOX ✲ BAYERITIAN ✲ BAYERTITAN ✲ BAYTITAN ✲ CALCOTONE WHITE T ✲ C.I. PIGMENT WHITE ✲ COSMETIC WHITE C47-5175 ✲ C-WEISS 7 (GERMAN) ✲ FLAMENCO ✲ HOMBITAN ✲ HORSE HEAD A-410 ✲ KRONOS TITANIUM DIOXIDE ✲ LEVANOX WHITE RKB ✲ RAYOX ✲ RUNA RH20 ✲ RUTILE ✲ TIOFINE ✲ TIOXIDE ✲ TITANIC ANHYDRIDE ✲ TITIANIC ACID ANHYDRIDE ✲ TITANIUM OXIDE ✲ TITIANIUM WHITE ✲ TITANIA ✲ TRIOXIDE(S) ✲ TRONOX ✲ UNITANE O-110 ✲ ZOPAQUE

TOLUENE

Products: In paint, solvents, lacquers, art supplies, nailpolish, cosmetics, dry cleaning products, spot removers, waxes, gasolines, detergents, dyes, pharmaceuticals, perfumes, and adhesive solvent for plastic toys and novelties.

Use: Derived from coal-tar. Used as a thinner and solvent. Various chemical uses.

Health Effects: Mildly toxic by breathing. Central Nervous System (CNS) effects on the human body by breathing are hallucinations, distorted perceptions, motor activity changes, antipsychotic, psychophysiological test changes, and bone marrow changes. An eye irritant.

Synonyms: CAS: 108-88-3 ✲ ANTISAL 1a ✲ BENZENE, METHYL- ✲ METHACIDE ✲ METHANE, PHENYL- ✲ METHYLBENZENE ✲ METHYLBENZOL ✲ PHENYLMETHANE ✲ TOLUEEN (DUTCH) ✲ TOLUEN (CZECH) ✲ TOLUOL (DOT) ✲ TOLUOLO (ITALIAN) ✲ TOLU-SOL

TONKA

Products: Derived from seeds of fruit of tree *Dipteyx Odorata*. In flavorings, extracts, powders, deodorants, detergents, shampoos, soaps, and antiseptics.

Use: As deodorizing and odor-enhancing agent; in pharmaceutical preparations.

Health Effects: Moderately toxic by swallowing inappropriate amounts. A skin irritant.

Synonyms: CAS: 91-64-5 ❊ COUMARIN ❊ CUMARIN ❊ BENZOPYRONE ❊ TONKA BEAN CAMPHOR

TRAGACANTH GUM

Products: In baked goods, beverages (citrus), condiments, fats, fruit fillings, gravies, meat products, oil, relishes, salad dressings, and sauces. In adhesives, leather dressing, toothpastes, soap chip coating, soap powders, hairwave preparations, printing inks, and medication tablet binders (holds components together).

Use: As an emulsifier (stabilizes and maintains mixes), preservative, and thickening agent.

Health Effects: Mildly toxic by swallowing excessive quantities of pure chemical. A mild allergen. FDA states GRAS (generally regarded as safe) when used within stated limits.

Synonyms: CAS: 9000-65-1 ❊ GUM TRAGACANTH ❊ TRAGACANTH

TRICHLOROETHYLENE

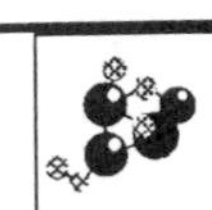

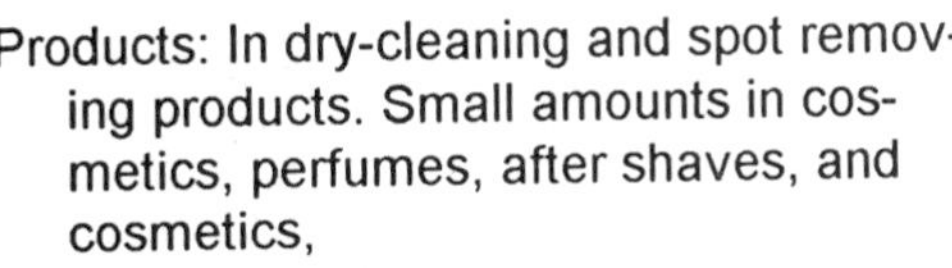

Products: In dry-cleaning and spot removing products. Small amounts in cosmetics, perfumes, after shaves, and cosmetics,

Use: As solvent, dye, adhesive, paint, and cleaner.

Health Effects: A suspected carcinogen (causes cancer). Toxic by breathing and swallowing. Effects on the body by swallowing are sleepiness, hallucinations, vision problems, confusion, gastrointestinal (stomach) problems, and jaundice (yellowing of skin and eyes). A human mutagen (changes inherited characteristics). A severe skin and eye irritant. Breathing causes headache, drowsiness, and possibly unconsciousness. Can cause cardiac failure.

Synonyms: CAS: 79-01-6 ❊ ACETYLENE TRICHLORIDE ❊ BENZINOL ❊ ETHINYL TRICHLORIDE ❊ EHYLENE TRICHLORIDE ❊ TRICHLORAN ❊ TRICHLORETHENE ❊ THRETHYLENE

TRICHLOROFLUORMETHANE

Products: For fire extinguishers, refrigerators, and solvents.

Use: An aerosol propellant and refrigerant.

Health Effects: High concentrations cause numbness and anesthesia. Effect on the body by breathing include eye irritation, lung and liver changes.

Synonyms: CAS: 75-69-4 ❊ FLUOROTRICHLOROMETHANE ❊ FLUOROCARBON-11

TRICLOSAN

Products: In deodorants, hand soaps, vaginal deodorants, cosmetics, pharmaceuticals, and cleansers.

Use: A disinfectant/antibacterial common in consumer toiletries.

Health Effects: Moderately toxic by swallowing. Mildly toxic by prolonged skin contact. A skin irritant and possible allergen.

Synonyms: CAS: 3380-34-5 ❊ IRGASAN ❊ 5-CHLORO-2-(2,4-DICHLOROPHENOXYPHENOL) ❊ 2′-HYDROXY-2.4.4′-TRICHLORO-PHENETHYLETHER

TRIETHANOLAMINE

Products: In drycleaning soaps, shaving soaps, shampoos, cosmetics, household detergents, fruit coatings, and vegetable coatings.

Use: A softening agent, emulsifier (stabilizes and maintains mixes), and humectant (keeps product from drying out).

Health Effects: Mildly toxic by swallowing. A skin and eye irritant.

Synonyms: CAS: 102-71-6 ❊ TRIHYDROXYTRIETHYLAMINE ❊ STEROLAMIDE ❊ TRIETHYLOLAMINE ❊ TROLAMINE ❊ DALTOGEN ❊ TEA

TRIPALMITIN

Products: In shampoos, soaps, shaving creams, and leather polish.

Use: As a texturizer (smooths ingredients) in toiletries and consumer products.

Health Effects: Harmless when used for intended purposes.

Synonyms: CAS: 555-44-2 ❊ PALMITIN ❊ GLYCERYL TRIPALMITATE ❊ HEXADECANOIC ACID

TRIPOLI

Products: In scouring soaps, scouring powders, polishing powder, paints, and wood filler.

Use: As an abrasive, absorbent, polish, filler, base, and filter.

Health Effects: The prolonged breathing of dusts may cause disabling lung problem know as silicosis.

Synonyms: CAS: 1317-95-9 ❊ ROTTENSTONE ❊ AMORPHOUS SILICA ❊ SILICA

TRISODIUM PHOSPHATE

Products: In water softeners, detergents, dishwashing compounds, and scouring powders.

Use: As a conditioner, cleaner, and bactericide.

Health Effects: Irritant to skin and eyes.

Synonyms: CAS: 7601-54-9 ❊ SODIUM PHOSPHATE, TRIBASIC ❊ PHOSPHORIC ACID, TRISODIUM SALT ❊ SODIUM PHOSPHATE ❊ TRISODIUM ORTHOPHOSPHATE ❊ TRISODIUM PHOSPHATE ❊ TROMETE

TRISULFURATED PHOSPHORUS

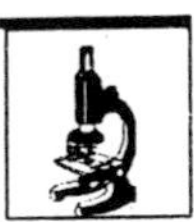

Products: In match tips.

Use: As a flammable solid.

Health Effects: Poison by swallowing.

Synonyms: CAS: 1314-85-8 ❊ TETRAPHOSPHORUS TRISULFIDE ❊ PHOSPHORUS SESQUISULFIDE

TRICHLOROISOCYANIC ACID

Products: In household dry bleaches, dishwashing compounds, scouring powders, detergent sanitizers, laundry bleaches, swimming pool, hot tub, and jacuzzi disinfectants

Use: As a bactericide, algicide, bleach, and deodorant.

Health Effects: Moderately toxic by swallowing. Effects on the body from swallowing include ulceration or bleeding from the stomach. A severe skin and eye irritant.

Synonyms: CAS: 87-90-1 ❊ ISOCYANIC CHLORIDE ❊ SYMCLOSEN ❊ TRICHLORINATED ISOCYANURIC ACID ❊ TRICHLOROISOCYANURIC ACID

TROMETHAMINE

Products: In leather polishes, cleaning polishes, skin lotions, pharmaceuticals, cosmetics, and creams.

Use: An emulsifier (stabilizes and maintains mixes) in polishes and cleaning compounds.

Health Effects: Moderately toxic by swallowing.

Synonyms: CAS: 77-86-1 ❊ THAM ❊ PEHANORM ❊ TALATROL ❊ TRIMETHYLOL AMINOMETHANE ❊ TRISAMINE ❊ TRIS BUFFER ❊ TROMETAMOL ❊ TROMETHANE ❊ TALALTROL

TSPP

Products: In shampoos, cleaning compounds, water softeners, rust stain remover, spot remover, and ink removers.

Use: As a sequestrant (binds ingredients that affect the final products appearance, flavor or texture), clarifying agent, and buffering (regulates, acidity and alkalinity) agent.

Health Effects: Poison by swallowing.

Synonyms: CAS: 7722-88-5 ✲ PYROPHOSPHATE ✲ TETRASODIUM PYROPHOSPHATE ✲ SODIUM PYROPHOSPHATE ✲ TETRASODIUM DIPHOSPHATE ✲ PHOSPHOTEX

TUNG NUT OIL

Products: Exterior paint and varnishes.

Use: Derived from seeds of plant grown in the Orient. Used for waterproofing, finishes, and packaging materials.

Health Effects: A skin irritant. Toxic by swallowing. Swallowing causes nausea, vomiting, cramps, diarrhea, rectal spasms, thirst, dizziness, sleepiness, and confusion. Large doses can cause fever, disturbance of heart rhythms, and breathing effects.

Synonyms: CHINAWOOD OIL

TURPENTINE

Products: In varnishes, insecticides, paint thinners, wax-based polishes, shoe polishes, furniture polishes, liniments (medicinal), antiseptics, and perfumery.

Use: A common solvent.

Health Effects: Moderately toxic by swallowing. Effects on the human body by swallowing and breathing are nasal and eye irritation, hallucinations, confusion, headache, lung and kidney damage. Irritating to skin, nose, and throats. A very dangerous fire hazard when exposed to heat or flames.

Synonyms: CAS: 8006-64-2 ✲ OIL OF TURPENTINE ✲ SPIRIT OF TURPENTINE ✲ TURPENTINE OIL

> "High alcohol mouthwashes may raise a person's risk of developing oral cancer by as much as 60%"
>
> *National Cancer Institute*

ULTRAMARINE BLUE

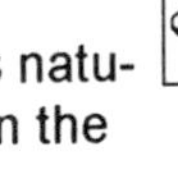

Products: As laundry bluing to offset the yellow tones. Also in soaps and cosmetics (eye shadows and mascaras).

Use: As a coloring agent. It occurs naturally as the blue pigmentation in the mineral lapis lazuli.

Health Effects: Not harmful when used for external use.

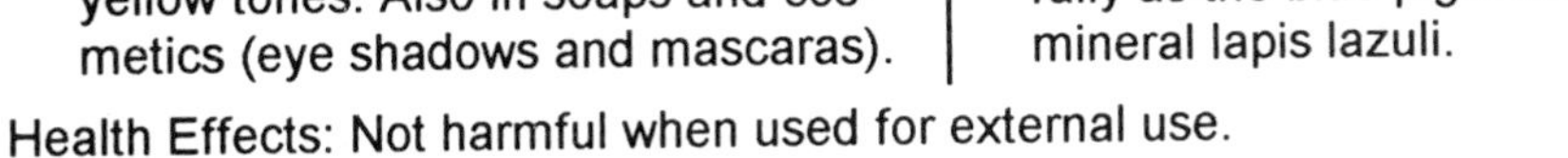

Synonyms: CAS: 57455-37-5 ✲ C.I. PIGMENT BLUE

UMBELLIFERONE

Products: Found in cosmetics and sunscreen cream, lotion, gel, spray preparations.

Use: As an ultraviolet (UV) screening agent.

Health Effects: Harmless when used for intended purposes.

Synonyms: CAS: 93-35-6 ✲ 7-HYDROXYCOUMARIN ✲ HYDRANGIN ✲ SKIMMETIN

γ-UNDECALACTONE

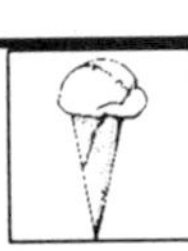

Products: In beverages, candy, gelatins, ice cream, puddings, and perfumes.

Use: An ingredient used to affect the taste of smell of the product. Provides a peach-like odor.

Health Effects: Harmless when used for intended purposes.

Synonyms: ALDEHYDE C-14 PURE ✲ FEMA No. 3091 ✲ PEACH ALDEHYDE

UNDECANAL

Products: In desserts, ice creams, bakery products, candies, and chewing gum.

Use: For fruit and floral flavorings and scents.

Health Effects: Chemical is a skin irritant in concentrated amounts.

Synonyms: CAS: 112-44-7 ✲ ALDEHYDE-14 ✲ 1-DECYL ALDEHYDE ✲ HENDECANAL ✲ HENDECANALDEHYDE ✲ UNDECANAL ✲ n-UNDECANAL ✲ UNDECANALDEHYDE ✲ UNDECYL ALDEHYDE ✲ N-UNDECYL ALDEHYDE ✲ UNDECYLIC ALDEHYDE

UNDECYL ALCOHOL

Products: In beverages, ice creams, candies, and bakery products.

Use: As a synthetic citrus and floral flavoring.

Health Effects: Moderately toxic by swallowing excessive amounts of pure chemical. A skin irritant.

Synonyms: CAS: 112-42-5 ✲ ALCOHOL C-11 ✲ HENDECANOIC ALCOHOL ✲ 1-HENDECANOL ✲ HENDECYL ALCOHOL ✲ n-HENDECYLENIC ALCOHOL ✲ n-UNDECANOL

UREA

Products: In alcoholic beverages, gelatin products, wine, yeast-raised bakery products for “browning”, plant fertilizer, and animal feeds. In personal care products, deodorants, toothpastes, mouthwashes, and hair products.

Use: As a fermentation aid, formulation aid, antiseptic, adhesive, and flameproofer.

Health Effects: Concentrated chemical is a human mutagen (changes inherited characteristics). A skin irritant. GRAS (generally regarded as safe) when used within limits.

Synonyms: CAS: 57-13-6 ✲ CARBAMIDE ✲ CARBAMIDE RESIN ✲ CARBAMIMIDIC ACID ✲ CARBONYL DIAMIDE ✲ CARBONYLDIAMINE ✲ ISOUREA ✲ PRESPERSION, 75 UREA ✲ PSEUDOUREA ✲ SUPERCEL 3000 ✲ UREAPHIL ✲ UREOPHIL ✲ UREVERT ✲ VARIOFORM II

UREA PEROXIDE

Products: In cosmetics and pharmaceuticals.

Use: As a bleaching disinfectant.

Health Effects: An irritant.

Synonyms: CAS: 124-43-6 ✲ HYDROGEN PEROXIDE CARBAMIDE ✲ EXTEROL ✲ ORTIZON ✲ HYPEROL ✲ PERHYDRIT ✲ PERHYDROL UREA

URETHANE

Products: Found in over 1000 beverages sold in the U. S. In liquors, bourbons, sherries, and fruit brandies. In whiskeys, table wines, dessert wines, brandies, liqueurs. A natural product of the

Use: As an intermediate in the manufacture of pharmaceuticals, pesticides and fungicides.

fermentation process by which yeast turns fruit juice into beverages. Allowable limit is 125 ppb. Some fruit brandies had 1000 to 12000 ppb urethane.

Health Effects: Pure substance is a definite carcinogen (causes cancer) that is toxic by swallowing. It is a human mutagen (changes inherited characteristics). Can cause CNS (central nervous system) depression, nausea, and vomiting.

Synonyms: CAS: 51-79-6 ❁ ETHYL CARBAMATE ❁ ETHYL URETHANE ❁ LEUCETHANE ❁ LEUCOTHANE ❁ PRACARBAMINE ❁ CARBAMIC ACID, ETHYL ESTER

URUSHIOL

Products: The oily, toxic components of poison ivy and poison oak.

Use: Sometimes used in hyposensitization therapy as an antiallergic.

Health Effects: Causes severe allergic dermatitis.

Synonyms: CATECHOL DERIVATIVE MIXTURE ❁ *RHUS TOXICODENDRON*

V

"An in-vitro (non-animal) test has been devised for testing chemical corrosivity and may be used to replace the cruel Draize (chemicals in eyes of rabbits) test in the future."

In-Vitro International (IVI)

VALERIAN OIL

Products: As a tobacco perfume, industrial odorant, and flavoring.

Use: Various.

Health Effects: Harmless when used for intended purposes.

Synonyms: PINENE ✲ CAMPHENE ✲ BORNEOL ✲ *VALERIANA OFFICINALIS*

VALERIC ACID

Products: In perfumes, pharmaceuticals, beverages, ice creams, candies, and bakery products.

Use: As an ingredient that affects the taste or smell of the product.

Health Effects: Moderately toxic by swallowing and breathing concentrated acid. A corrosive irritant to skin, eyes, nose, and throats.

Synonyms: CAS: 109-52-4 ✲ BUTANECARBOXYLIC ACID ✲ 1-BUTANECARBOXYLIC ACID ✲ PENTANOIC ACID ✲ n-PENTANOIC ACID ✲ PROPYLACETIC ACID ✲ VALERIANIC ACID ✲ n-VALERIC ACID

VANILLIN

Products: In perfumes, candies, beverages, various foods, liqueurs, and pharmaceuticals.

Use: As an ingredient that affects the aroma, taste, or scent, of final product.

Health Effects: Pure substance is moderately toxic by swallowing excessive amounts. FDA approves use in moderate levels.

Synonyms: CAS: 121-33-5 ✲ 4-HYDROXY-m-ANISALDEHYDE ✲ 4-HYDROXY-3-METHOXYBENZALDEHYDE ✲ LIOXIN ✲ 3-METHOXY-4-HYDROXYBENZALDEHYDE ✲ METHYLPROTOCATECHUALDEHYDE ✲ VANILLA ✲ VANILLALDEHYDE ✲ VANILLIC ALDEHYDE ✲ p-VANILLIN ✲ ZIMCO

VEGETABLE OIL

Products: In paints, shortenings, salad

Use: As a softener, carrier, filler,

dressings, margarine, soaps, cosmetics, lipstick, hair products, and shaving products.

thickener, or a cleanser.

Health Effects: Harmless when used for intended purposes. Some individuals may have allergic reactions as the oil is derived from plant products such as peanut, olive, coconut, linseed, sesame, or cottonseed oil.

Synonyms: SEED OIL ❋ NUT OIL ❋ PLANT OIL ❋ MIXED GLYCERIDES

VERMICULITE

Products: For insulation, sound conditioning, fertilizer additive, plaster, soil conditioner, seed bed for plants, and absorbent for oil spills in the ocean.

Use: As filler, nonconductor, aerator, sponge, gypsum board, and leavening material.

Health Effects: Considered harmless when used for intended purposes.

Synonyms: HYDRATED MAGNESIUM-IRON-ALUMINUM SILICATE

VITAMIN A

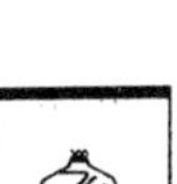

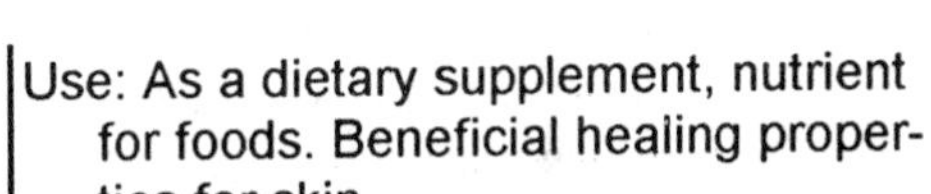

Products: In milk, cheese, margarine, ice cream, and baby formulations. Also in skin ointments and creams.

Use: As a dietary supplement, nutrient for foods. Beneficial healing properties for skin.

Health Effects: Moderately toxic by swallowing excessive quantities of pure chemical substance. In animals it has possible teratogenic (abnormal fetus development) effects by swallowing: developmental abnormalities of the craniofacial area, and urogenital system. A human mutagen (changes inherited characteristics). FDA states GRAS (generally regarded as safe) when used at moderate levels to accomplish the intended effect.

Synonyms: CAS: 68-26-8 ❋ AFAXIN ❋ AGIOLAN ❋ ALPHASTEROL ❋ ANATOLA ❋ ANTI-INFECTIVE VITAMIN ❋ ANTIXEROPHTHALMIC VITAMIN ❋ AORAL ❋ APEXOL ❋ AVIBON ❋ AVITOL ❋ BIOSTEROL ❋ CHOCOLA A ❋ 3,7-DIMETHYL-9-(2,6,6-TRIMETHYL-1-CYCLOHEXEN-1-YL)-2,4,6,8-NONATETRAEN-1-OL ❋ DISATABS TABS ❋ DOFSOL ❋ EPITELIOL ❋ HI-A-VITA ❋ LARD FACTOR ❋ MYVPACK ❋ OLEOVITAMIN A ❋ OPHTHALAMIN ❋ PREPALIN ❋ RETINOL ❋ all-trans RETINOL ❋ RETROVITAMIN A ❋ TESTAVOL ❋ VAFLOL ❋ VI-ALPHA ❋ VITAMIN A1 ❋ VITAMIN A1 ALCOHOL ❋ all-trans-VITAMIN A ALCOHOL ❋ VITAVEL-A ❋ VITPEX ❋ VOGAN

VITAMIN E

Products: In bacon, fats (rendered animal), pork fat (rendered), and poultry.

Use: As an antioxidant (added to oil-containing food to prevent it from getting rancid), dietary supplement, nutrient, or preservative.

Health Effects: FDA states GRAS (generally regarded as safe) when used within stated limits.

Synonyms: CAS: 59-02-9 ❋ ALMEFROL ❋ ANTISTERILITY VITAMIN ❋ COVI-OX ❋ DENAMONE ❋ EMIPHEROL ❋ ENDO E ❋ EPHYNAL ❋ EPROLIN ❋ EPSILAN ❋ ESORB ❋ ETAMICAN ❋ ETAVIT ❋ EVION ❋ EVITAMINUM ❋ ILITIA ❋ PHYTOGERMINE ❋ PROFECUNDIN ❋ SPAVIT ❋ SYNTOPHEROL ❋ *d*-α-TOCOPHEROL (FCC) ❋ *dl*-α-TOCOPHEROL (FCC) ❋ (R,R,R)-α-TOCOPHEROL ❋ α-TOCOPHEROL ❋ (2R,4′R,8′R)-α-TOCOPHEROL ❋ TOKOPHARM ❋ 5,7,8-TRIMETHYLTOCOL ❋ VASCUALS ❋ VERROL ❋ VITAPLEX E ❋ VITAYONON ❋ VITEOLIN

VOLATILE ORGANIC COMPOUNDS

Products: In hair sprays, windshield glass cleaner, air fresheners, liquid cleaning products, auto engine degreasers, wood furniture polish, wood floor polish, laundry products, nail polish remover, oven cleaners, hair mousse spray, bath/ceramic cleaners, spray insect repellents, hair setting gels, and shaving creams. Emitted from carpeting, furniture, subflooring, dry-cleaned clothes, curtains, and the wall material in mobile homes. Permanent press clothes are treated with formaldehyde when they are made.

Use: Solvent, vehicle.

Health Effects: Irritation of the eyes, nose and throat. Headaches, mental fatigue, and breathing difficulties.

Synonyms: VOC ❋ VOLATILE ORGANIC CHEMICALS ❋ FORMALDEHYDE ❋ BENZENE ❋ TCE ❋ ORGANIC VOLATILE CHEMICALS ❋ OVC

"All women of childbearing age require daily amounts of folic acid, a B vitamin, which prevents infant brain and spinal cord defects. It's found in spinach, broccoli, or folic acid supplements."

United States Public Health Service

WARFARIN

Products: Rodenticide, and medical anticoagulant.

Use: Rat and mouse poison; and oral medications.

Health Effects: A poison by swallowing and breathing. Moderately toxic by skin contact. Effects on the human body by swallowing include hemorrhage, ulceration, bleeding from small intestine. Human reproductive effects by swallowing include fetus death and abnormalities at birth. Human teratogenic (abnormal fetal development) effects include abnormal head, face, musculoskeletal, and breathing problems.

Synonyms: CAS: 81-81-2 ✲ ARAB RAT DETH ✲ BRUMIN ✲ COMPOUND 42 ✲ d-CON ✲ CO-RAX ✲ COUMADIN ✲ COUMAFENE ✲ LIQUA-TOX ✲ PROTHROMADIN ✲ RAT-A-WAY ✲ RAT-B-GON ✲ RAT-GARD ✲ SOLFARIN

WASHING SODA

Products: In soaps, detergents, cleaning, and bleaching preparations.

Use: As cleanser and sanitizer.

Health Effects: Moderately toxic by swallowing or breathing. A skin and eye irritant.

Synonyms: CAS: 497-19-8 ✲ CARBONIC ACID ✲ DISODIUM SALT ✲ CRYSTOL CARBONATE ✲ DISODIUM CARBONATE ✲ SODA ASH ✲ TRONA

WAX

Products: In polishes, carbon paper, floor wax, candles, crayons, sealants, cosmetics, and for protecting food products.

Use: Various.

Health Effects: Harmless when used for intended purposes.

Synonyms: WAX, CHLORONAPHTHALENE ✲ WAX, MICROCRYSTALLINE ✲ WAX, POLYMETHYLENE

WHEAT GLUTEN

Products: In bakery bread items. Makeup powders and creams.

Use: As dough conditioner, formulation aid, nutrient supplement, processing aid, stabilizer, surface-finishing agent, texturizing agent, and thickening agent.

Health Effects: FDA states GRAS (generally regarded as safe).

Synonyms: CAS: 8002-80-0

WHEY, DRY

Products: Beef with barbecue sauce, bockwurst, chili con carne, loaves (nonspecific), pork with barbecue sauce, poultry, sausage, sausage (imitation), soups, and stews.

Use: As a binder or extender.

Health Effects: FDA states GRAS (generally regarded as safe) when used within stated limits.

Synonyms: DRY WHEY ❊ DRIED WHEY

WHISKEY

Products: As beverage or medicine.

Use: As stimulant, antiseptic, or vasodilator.

Health Effects: A noncumulative poison usually harmless in moderate amounts, but may be toxic when habitually taken in large amounts.

Synonyms: CORN WHISKEY ❊ RYE WHISKEY ❊ BARLEY WHISKEY ❊ ALCOHOL ❊ ETHANOL

WHITEWASH

Products: Wall coatings and paint-type products.

Use: As antiseptic coating for barns and chicken houses

Health Effects: Package directions should be carefully observed using necessary eye and skin protection.

Synonyms: KALSOMINE ❊ HYDRATED LIME SUSPENSION ❊ CALCIUM CARBONATE SUSPENSION.

WILD CHERRY

Products: Bark and stems used in food, cosmetics, pharmaceuticals, cough

Use: An ingredient that affects the taste or smell of final product.

drops, syrups, and expectorants.

Health Effects: Harmless when used for intended purposes.

Synonyms: *PRUNUS SEROTINA*

WINE

Products: The fermented juice of grapes or other fruits or plants.

Use: Beverages.

Health Effects: Some studies have indicated that small quantities (four ounces per day) have been found to be beneficial in the avoidance of heart attacks. Large quantities over extended periods can lead to alcoholism, liver damage, numbness of the extremities, brain damage, gastritis, and heart muscle damage. Pregnant women should avoid alcohol entirely because of the disastrous effects on the unborn.

Synonyms: ALCOHOLIC BEVERAGE

WOODRUFF

Products: In wine and sachets.

Use: An herb for flavoring May wine and for aromatic pot-pourri.

Health Effects: Harmless when used for intended purposes.

Synonyms: *ASPERULA ODORATA* LEAVES

X

"It is a myth that children who consume excessive amounts of sugar are hyperactive, studies show the opposite is true, they are less active."

Food Technology

XANTHAN GUM

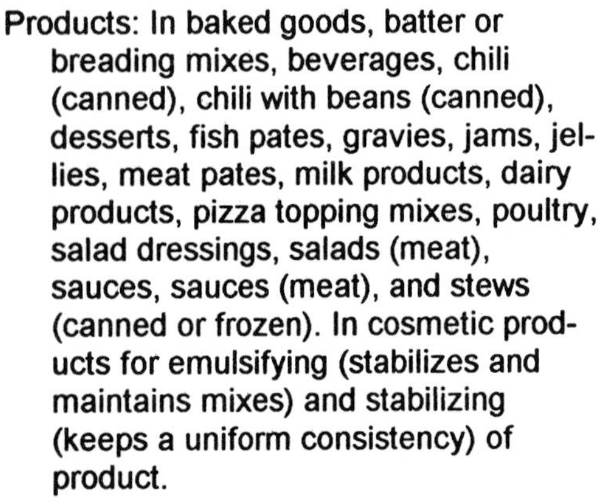

Products: In baked goods, batter or breading mixes, beverages, chili (canned), chili with beans (canned), desserts, fish pates, gravies, jams, jellies, meat pates, milk products, dairy products, pizza topping mixes, poultry, salad dressings, salads (meat), sauces, sauces (meat), and stews (canned or frozen). In cosmetic products for emulsifying (stabilizes and maintains mixes) and stabilizing (keeps a uniform consistency) of product.

Use: As a binder, bodying agent, emulsifier, extender, foam stabilizer, stabilizer, suspending agent, and thickening agent.

Health Effects: FDA approves use at moderate levels to accomplish the desired results. USDA states use limitations.

Synonyms: CAS: 11138-66-2 ✼ CORN SUGAR GUM ✼ *XANTHOMONAS CAMPESTRIS* ✼ POLYSACCHARIDE GUM

XYLENOL

Products: In disinfectants, lubricants, gasolines, solvents, pharmaceuticals, insecticides, and fungicides.

Use: Various.

Health Effects: Toxic by swallowing and skin absorption.

Synonyms: CAS: 1300-71-6 ✼ DIMETHYLPHENOL ✼ HYDROXYDIMETHYLBENZENE ✼ DIMETHYLHYDROXYBENZENE

XYLITOL

Products: In gum, candies, and breath

Use: A nutritive sweetener.

mints.

Health Effects: FDA approves use at moderate levels to accomplish the intended effect. Mildly toxic by swallowing excessive amounts of pure chemical. A sugar.

Synonyms: CAS: 87-99-0 ✲ KLINIT ✲ XYLITE (SUGAR)

"Studies have identified 300 toxic chemicals and 150 cancer-causing chemicals used in common paint formulations. It is necessary to ventilate well when using."

Johns Hopkins University

YEAST

Products: In baked, cooked, brewed food and beverage products.

Use: Fermentation of sugars, molasses, and cereals for alcohol; food supplement; source of vitamins.

Health Effects: Harmless when used for intended purposes.

Synonyms: BARM ✲ *SACCHAROMYCETACEAE*

YLANG YLANG OIL

Products: In fragrances, perfumes, and colognes.

Use: An ingredient that affects the aroma.

Health Effects: Harmless when used for intended purposes.

Synonyms: CANANGA OIL ✲ *CANANGA ODORATA*

Z

"Smoking is the number one cause of avoidable deaths in this country. Yet, there are still 46 million smokers."

American Council on Science and Health

ZANZIBAR GUM

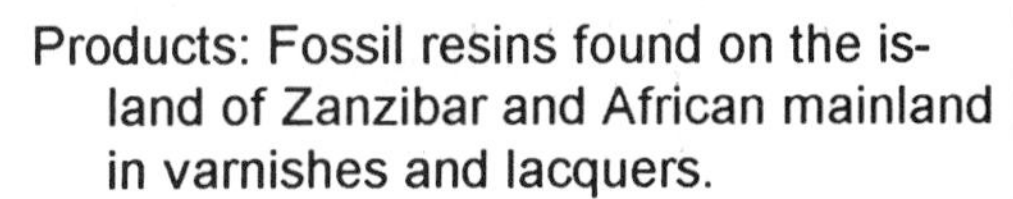

Products: Fossil resins found on the island of Zanzibar and African mainland in varnishes and lacquers.

Use: An ingredient in wood-finishing products.

Health Effects: Harmless when used for intended purposes. Label directions on products should be carefully followed.

Synonyms: AFRICAN GUM ✲ ZANZIBAR ISLAND GUM

ZEIN

Products: In confections, grain, nuts, adhesives, and panned goods.

Use: As a glaze, paper coating, grease-resistant coating, label varnishes, printing inks, shellac substitute, food coatings, microencapsulation, and surface-finishing agent.

Health Effects: A corn processing by-product considered non-toxic and harmless when used for intended purposes. FDA states GRAS (generally regarded as safe).

Synonyms: CAS: 9010-66-6

ZEOLITE

Products: Water softeners, detergent builders, adsorbents, and desiccants (drying agent).

Use: Various.

Health Effects: Container directions must be followed carefully.

Synonyms: ANALCITE ✲ CHABAZITE ✲ HEULANDITE ✲ NATROLITE ✲ STILBITE ✲ THOMOSONITE

ZINC ACETATE

Products: In astringents, antiseptics, emetics, and styptics. As wood preservative, dietary supplement, textile dyes, and feed additives.

Use: Various.

Health Effects: Harmless when used for intended purposes.

Synonyms: CAS: 557-34-6

ZINC ARSENATE

Products: An insecticide and wood preservative.

Use: Various.

Health Effects: A confirmed carcinogen (causes cancer). Toxic by swallowing and breathing.

Synonyms: ZINC ORTHOARSENATE ✲ KOETTIGITE MINERAL

ZINC BACITRACIN

Products: As antibacterial, antiseptic ointment, suppositories, throat lozenges, and other pharmaceuticals.

Use: In assorted medical products as an antimicrobial.

Health Effects: Harmless when used for intended purposes.

Synonyms: BACITRACIN ZINC COMPLEX ✲ BACIFERM

ZINC CARBONATE

Products: In cosmetics, lotions, pharmaceuticals, ointments, and dusting powders.

Use: As a coloring agent in toiletries and consumer products.

Health Effects: Harmless when used for intended purposes.

Synonyms: CAS: 3486-35-9 ✲ SMITHSONITE ✲ ZINCSPAR

ZINC CHLORIDE

Products: In adhesives, dental cements, deodorants, disinfectants, and taxidermy embalming fluid. On lumber it is used for a preservative and fireproofing. In medicines as an astringent and antiseptic.

Use: Various.

Health Effects: Poison by swallowing excessive concentrated amounts. Lung tissue changes can result from breathing. A corrosive irritant to the skin, eyes, nose, and throat.

Synonyms: CAS: 7646-85-7 ✲ BUTTER OF ZINC ✲ TINNING GLUX ✲ ZINC DICHLORIDE ✲ ZINC MURIATE

ZINC CITRATE

Products: In toothpaste, dentifrices, breath fresheners, and mouth washes.

Use: As an oral hygiene product.

Health Effects: Harmless when used for intended purposes.

Synonyms: ZINC CARBONATE CITRIC ACID

ZINC DITHIOAMINE COMPLEX

Products: Fungicide, rat and mouse poison, deer, and rabbit repellent.

Use: As a pesticide and wildlife deterrent.

Health Effects: Toxic by swallowing.

Synonyms: ZINC DIMETHYLDITHIOCARBAMATECYCLOHEXYLAMINE

ZINC FLUOROSILICATE

Products: In concrete hardener, laundry sour, preservative, and mothproofing agents.

Use: Various.

Health Effects: Poison by swallowing.

Synonyms: CAS: 16871-71-9 ✲ ZINC SILICOFLUORIDE ✲ ZINC FLUOSILICATE ✲ ZINC HEXAFLUOROSILICATE

ZINC FORMATE

Products: Waterproofing agents, textiles, and antiseptic.

Use: Various.

Health Effects: Toxic by swallowing.

Synonyms: CAS: 557-41-5

ZINC METHIONINE SULFATE

Products: In vitamin tablets.

Use: As a dietary supplement or nutrient.

Health Effects: Harmless when used for intended purposes.

Synonyms: CAS: 56329-42-1

ZINC OXIDE

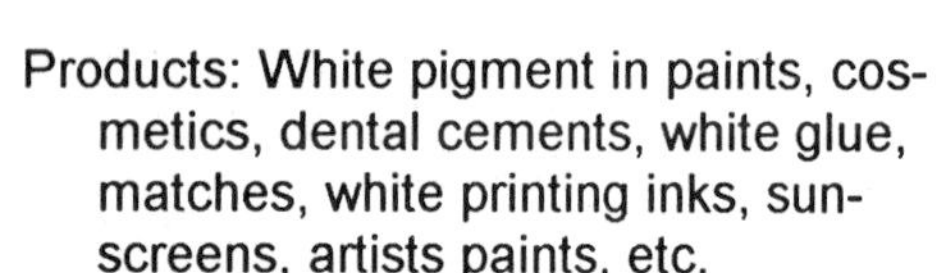

Products: White pigment in paints, cosmetics, dental cements, white glue, matches, white printing inks, sunscreens, artists paints, etc.

Use: As a coloring agent, antiseptic, astringent, dietary supplement, UV (ultraviolet) absorber, and nutrient.

Health Effects: Moderately toxic to humans by swallowing. A skin and eye irritant.

Synonyms: CAS: 1314-13-2 ⁂ AMALOX ⁂ AZODOX-55 ⁂ CALAMINE (spray) ⁂ CHINESE WHITE ⁂ EMANAY ZINC OXIDE ⁂ EMAR ⁂ FELLING ZINC OXIDE ⁂ FLOWERS of ZINC ⁂ HUBBUCK'S WHITE ⁂ K-ZINC ⁂ OZIDE ⁂ OZLO ⁂ PASCO ⁂ PERMANENT WHITE ⁂ PHILOSOPHER'S WOOL ⁂ SNOW WHITE ⁂ WHITE SEAL ⁂ ZINCITE ⁂ ZINCOID ⁂ ZINC WHITE

ZINC PROPIONATE

Products: For adhesive tape and medical bandaging.

Use: As a topical fungicide (kills fungi, molds, and bacteria).

Health Effects: Harmless when used for intended purposes.

Synonyms: CAS: 557-28-8

ZINC STEARATE

Products: In tablets, cosmetics, pharmaceuticals, powders, and ointments. As a waterproofing agent.

Use: In antiseptics, astringents, lacquers, and protective coating.

Health Effects: Breathing can cause lung damage.

Synonyms: CAS: 557-05-1

ZIRCONYL CHLORIDE

Products: In body deodorants, antiperspirants, cosmetic additives, and as water repellents.

Use: In topical skin products.

Health Effects: Moderately toxic by swallowing. Can cause skin irritation.

Synonyms: CAS: 7699-43-6 ⁂ ZIRCONIUM OXYCHLORIDE ⁂ BASIC ZIRCONIUM CHLORIDE ⁂ ZIRCONIUM CHLORIDE

APPENDIX I

Suggested Additional Readings and References

Berkow, Robert, M.D., ed. *The Merck Manual.* 15th ed. Rahway, New Jersey: Merck & Co., Inc. 1987.

Budavari, Susan, ed. *The Merck Index.* 11th ed. Rahway, New Jersey: Merck & Co., Inc. 1989.

Code of Federal Regulations. 21 CFR 1301.01. Washington, DC: U. S. Government Printing Office, 1992.

Edell, Dean, M.D., ed. *The Edell Health Letter.* San Francisco, California: Hippocrates Partners, 1992.

Edelston, Martin, ed. *Bottom Line/Personal.* New York, New York: Boardroom Reports, Inc. 1993.

Food Chemicals Codex. 3rd ed. National Research Council. Washington, DC: National Academy Press. 1981

Green, Nancy Sokol, *Poisoning Our Children.* Chicago, Illinois: The Noble Press, Inc. 1991

Hensyl, William R., ed. *Stedman's Medical Dictionary.* 24th ed. Baltimore, Maryland: Williams & Wilkins. 1982.

Igoe, Robert S., *Dictionary of Food Ingredients.* New York, New York: Van Nostrand Reinhold Co. 1983.

Jacobson, Michael F., *Safe Food.* Los Angeles, California: Center for Science in the Public Interest and Living Planet Press. 1991.

Lewis, Richard J., Sr. ed., *Dangerous Properties of Industrial Materials.* 8th ed. New York, New York: Van Nostrand Reinhold Co. 1992.

Lewis, Richard J., Sr. ed., *Food Additives Handbook.* New York, New York: Van Nostrand Reinhold Co. 1989.

Lewis, Richard J., Sr. ed., *Hawley's Condensed Chemical Dictionary*. 12th ed. New York, New York: Van Nostrand Reinhold Co. 1993.

Lewis, Richard J., Sr. ed., *Rapid Guide to Hazardous Chemicals in the Workplace*. 3rd ed. New York, New York: Van Nostrand Reinhold Co. 1993

Lippmann, Morton, PhD., Ed. *Environmental Toxicants: Human Exposures and Their Health Effects*. New York: Van Nostrand Reinhold Co. 1992

Ottoboni, M. Alice. *The Dose Makes the Poison,* Second Edition. New York: Van Nostrand Reinhold Co. 1991

Saulson, Donald, and Elisabeth Saulson. *A Pocket Guide to Food Additives*. Huntington Beach, California: VPS Publishing. 1991

Sweet, Doris V., ed. *The Registry of Toxic Effects of Chemical Substances*, Washington, DC: U. S. Government Printing Office, 1992.

Pohanish, Richard P., and Stanley A. Greene. *Hazardous Substances Resource Guide.* Detroit: Gale Research Inc. 1993.

APPENDIX II

Glossary

Acaracide - A tick and mite killer.
Allergen - A material that makes the body produce an antibody that causes allergy in hypersensitive individuals.
Amino acids - The building blocks of proteins. Their are 22 and 8 must be supplied by diet as they are not produced by body.
Anoxia - Refers to the absence of oxygen in blood or tissues.
Antipruritic - An anti-itching agent.
Antipyretic - An agent used to reduce fever.
Antirachitic - An agent that cures rickets.
Asphyxiant - Something that can cause smothering or suffocation or inability to breathe.
Ataxia - Refers to lack of coordination.
Ataxia - The loss of muscle coordination.
Auditory - Relating to hearing.
BATF - Bureau of Alcohol, Tax and Firearms.
Botulism - Poison resulting from improperly preserved or canned foods.
Central Nervous System Depressants - A substance that slows the heart rate or breathing rate.
Coma - A deep unconsciousness.
Coma - A state of unconsciousness from which one cannot be roused.
Comodgenic - A substance that clogs the pores of the skin.
Cyanosis - A bluish coloration of the skin due to lack of oxygen.
Demulcent - An agent that relieves skin irritation.
Dermatitis - Refers to skin irritation or inflammation.
Diaphoresis - Perspiration.
Dyspepsia - Indigestion or upset stomach.
Dyspnea - Difficult breathing and/or shortness of breath.
EPA - Environmental Protection Agency
Eczema - Skin condition exhibited by redness, inflammation, scaling, itching, burning, among others.
Edema - Refers to accumulation of excessive fluid in tissues.
Emphysema - Breathlessness caused by reduction of alveolar surfaces. Condition frequently coexists with chronic bronchitis.
Euphoria - A feeling of exaggerated well-being.
FDA - Food and Drug Administration
Gastritis - Inflammation of the (mucosal) stomach.
Gastrointestinal - Relating to the stomach and intestines.
GRAS - Generally Regarded As Safe. A list established in 1958 by Food and Drug Administration. The list changes depending upon new research and regulations.
Guano - An excrement frequently used as fertilizer.
Hematuria - Refers to blood in the urine.
Hematuria - Refers to blood in urine.
Intravenous - Refers to injection in vein.
Irritant - An agent that causes inflammation or irritation to skin or other organ.

Jaundice - Yellow coloration of eyes or skin caused by excessive billirubin.
Lachrymator - A substance that causes excessive eye watering or tearing.
Lacrimation - Refers to excessive watering or tearing of the eyes.
Lethargy - Refers to tired feeling, fatigue, or lack of energy.
Lymphoma - A general term for malignant tumor.
Malaise - A feeling of sickness.
Microencapsulation - Medication, fertilizer, etc., enclosed in a tiny capsule or sheath that dissolves when moistened.
Mutagenic - A change in the characteristics of a gene that will be reproduced in later cell divisions. A substance that changes inherited characteristics.
Narcosis - A stupor, drowsiness, or unconsciousness.
Nematocide - Kills parasitic worms.
Olfactory - Related to nose and sense of smell.
Outgassing - The release of chemicals in the form of gas or fumes from solids.
Oxidizer - A substance that yields oxygen.
Pallor - A lack of color in skin; a paleness of skin.
Palpitate - To throb or pulse with an increased frequency.
Paresthesia - Refers to abnormal sensations of burning or tingling.
Pediculicide - An agent that is used to kill lice.
Perceptual disturbance - The inability to judge distance.
Peripheral - Refers to the outside edge; the opposite of center.
Pharmacologic effect - The result of chemical action.
Pulmonary - Related to lungs and breathing.
Renal - Related to the kidneys.
Sensitizer - A material that causes many people to develop an allergic reaction after multiple exposures.
Sizing - Smooth or stiff finish on clothing or fabrics.
Solvent - A chemical liquid that dissolves and in some cases aids in the mixing of some other product.
Somnolence - Refers to sleepiness or drowsiness.
Somnolence - Sleepiness.
Subcutaneous - To be injected under the skin.
Teratogenic - Something that causes abnormal fetal development.
Tinnitus - Ringing in the ears.
Toxic - Something that is poison. Reactions range from slightly ill to death.
USDA - United States Department of Agriculture
Vascular - Relating to blood vessel.
Vertigo - Dizziness.

APPENDIX III

Poison Control Centers

NATIONWIDE
National Animal Poison Control Center
University of Illinois
1-800-548-2423 (fee)
1-900-680-0000 (per minute charge)

ALABAMA
Alabama Poison Control Center
1-205-345-0600
1-800-462-0800

Alabama Children's Hospital Poison Control Center
1-205-939-9201
1-205-933-4050
1-800-292-6678

ALASKA
Anchorage Alaska Poison Center
1-800-478-3193

ARIZONA
Arizona Poison and Drug Information Center
1-602-626-6016
1-800-362-0101

Phoenix Arizona Samaritan Regional Poison Center
1-602-253-3334

ARKANSAS
Arkansas Statewide Poison Control Drug Information Center
1-800-482-8948

CALIFORNIA
Fresno Regional Poison Control Center of Fresno Community Hospital and Medical Center
1-209-445-1222
1-800-346-5922

Los Angeles County Medical Association Regional Poison Center
1-213-484-5151
1-800-825-2722
1-800-777-6476

San Diego Regional Poison Center
1-619-543-6000
1-800-876-4766

San Francisco Bay Area Regional Poison Control Center
1-415-476-6600
1-800-523-2222

Santa Clara Valley Medical Center Regional Poison Center
1-800-662-9886

USDMC Regional Poison Control Center
1-916-453-3414
1-800-342-9293

COLORADO
Mid-Plains Poison Control Center
1-402-390-5400
1-800-642-9999
1-800228-9515

Rocky Mountain Poison and Drug Center
1-303-629-1123
1-800-332-3073
1-800-525-5042
1-800-442-2702

DISTRICT OF COLUMBIA
National Capital Poison Center
1-202-625-3333
1-202-784-4660

FLORIDA
Florida Poison Information Center at the Tampa General Hospital
1-813-253-4444
1-800-282-3171

GEORGIA
Georgia Poison Control Center
1-404-589-4400
1-404-525-3323
1-800-282-5846

HAWAII
Hawaii Poison Center
1-800-362-3585

IDAHO
Idaho Poison Control Center
1-800-632-8000
Spokane Poison Center
1-800-572-5842

ILLINOIS
Central and Southern Illinois Regional Poison Resource Center
1-800-252-2022

Chicago and Northeastern Illinois Regional Poison Control Center
1-800-942-5969

INDIANA
Indiana Poison Center
1-317-929-2323
1-800-382-9097

IOWA
McKennen Hospital Poison Center
1-800-952-0123
1-800-843-0505

Mid-Plains Poison Control Center
1-402-390-5400
1-800-642-9999
1-800-228-9515

University of Iowa Hospitals and Clinics Poison Control Center
1-800-272-6477
Variety Club Poison and Drug Information Center
1-800-362-2327

KANSAS
Mid-American Poison Center
1-800-322-6633
Mid-Plains Poison Control Center
1-402-390-5400
1-800-642-9999
1-800-228-9515

KENTUCKY
Kentucky Regional Poison Control Center of Kosair Children's Hospital
1-502-589-8222
1-800-722-5725

MAINE
Maine Poison Control Center at Maine Medical Center
1-800-442-6305

MARYLAND
Maryland Poison Center
1-301-528-7701
1-800-492-2414

National Capital Poison Center
1-202-625-3333
1-202-784-4660

MASSACHUSETTS
1-617-232-2120
1-800-682-9211

MICHIGAN
Blodgett Regional Poison Center
1-800-632-2727
1-800-356-3232

Children's Hospital of Michigan Poison Control Center
1-313-745-5711
1-800-462-6642

Great Lakes Poison Center
1-800-442-4112 (within area code 616)

MINNESOTA
Hennepin Regional Poison Center
1-612-347-3141
1-612-337-7474

McKennen Hospital Poison Center
1-800-952-0123
1-800-843-0505 (in Nebraska, Iowa, and Minnesota)

Minnesota Regional Poison Center
1-612-221-2113
1-800-222-1222

MISSOURI
Cardinal Glennon Children's Hospital Regional Poison Center
1-314-772-5200
1-314-577-5336
1-800-392-9111
1-800-366-8888

MONTANA
Rocky Mountain Poison and Drug Center
1-303-629-1123
1-800-332-3073
1-800-525-5042 (in Montana)
1-800-442-2702 (in Wyoming)

NEBRASKA
McKennon Hospital Poison Center
1-800-952-0123
1-800-843-0505 (in Iowa, Minnesota, Nebraska)

Mid-Plains Poison Control Center
1-402-390-5400
1-800-642-9999-1-800-228-9515 (in Colorado, Iowa, Kansas, Missouri, South Dakota, Wyoming)

The Poison Center
1-402-390-5400
1-800-642-9999
1-800-228-9515

NEW HAMPSHIRE
New Hampshire Poison Information Center
1-800-562-8236

NEW JERSEY
New Jersey Poison Information and Education System
1-201-923-0764
1-800-962-1253

NEW MEXICO
New Mexico Poison and Drug Information Center
1-505-843-2551
1-800-432-6866

NEW YORK
Central New York Poison Control Center
1-800-252-5655

Long Island Regional Poison Control Center
1-516-542-2323

New York Poison Control Center
1-212-340-4494

NORTH CAROLINA
Duke University Poison Control Center
1-800-672-1697

Triad Poison Center
1-800-722-2222 (in North Carolina)

Western North Carolina Poison Control Center
1-800-542-4225

NORTH DAKOTA
North Dakota Poison Information Center
1-800-732-2200

OHIO
Akron Regional Poison Control Center
1-800-362-9922 (in Ohio)

Central Ohio Poison Control Center
1-614-228-1323
1-614-228-2272
1-800-682-7625 (in Ohio)

Lorain Community Hospital Poison Control Center
1-800-821-8972

Regional Poison Control System and Cincinnati Drug and Poison Information Center
1-513-558-5111
1-800-872-5111

Western Ohio Regional Poison and Drug Information Center
1-800-762-0727

OKLAHOMA
Oklahoma Poison Control Center
1-800-522-4611

OREGON
Oregon Poison Center
1-503-279-8968
1-800-452-7165

PENNSYLVANIA
Delaware Valley Regional Poison Control Center
1-215-386-2100

Northwest Regional Poison Center
1-800-822-3232

Pittsburgh Poison Center
1-412-681-6669

RHODE ISLAND
Rhode Island Poison Center
1-401-277-5727

SOUTH DAKOTA
McKennen Hospital Poison Center
1-800-952-0123
1-800-843-0505

Mid-Plains Poison Control Center
1-402-390-5400
1-800-228-9515

St. Luke's Poison Control Center
1-800-592-1889

TEXAS
North Texas Poison Center
1-214-590-5000
1-800-441-0040

Texas State Poison Center
1-409-765-1420
1-713-654-1701
1-512-478-4490
1-800-392-8548 (in Texas)

UTAH
Intermountain Regional Poison Control Center
1-801-581-2151
1-800-456-7707 (in Utah)

VIRGINIA
BlueRidge Poison Center
1-800-451-1428

National Capital Poison Center
1-202-625-3333
1-202-784-4660

Tidewater Poison Center
1-800-522-6337

WASHINGTON
1-800-572-9176 (in Washington)

Mary Bridge Poison Center
1-800-542-6319 (in Washington)

Seattle Poison Center
1-800-732-6985

Spokane Poison Center
1-800-572-5842
1-800-541-5624 (in Northern Idaho and western Wyoming)

WEST VIRGINIA
West Virginia Poison Center
1-304-348-4211
1-800-642-3625 (in West Virginia)

WYOMING
Mid-Plains Poison Control Center
1-402-390-5400
1-800-642-9999

1-800-228-9515 (in Colorado, Iowa, Kansas, Missouri, South Dakota, Wyoming)

Rocky Mountain Poison and Drug Center
1-303-629-1123
1-800-332-3073
1-800-525-5042 (in Montana)

1-800-442-2702 (in Wyoming)

SPOKANE POISON CENTER
1-800-572-5842
1-800-541-5624 (in No. Idaho and western Wyoming)

INDEX

The terms in bold type refer to the more familiar words while the smaller-sized type refers to the chemical names and synonyms.

A

B

C

D

E

F

G

H

I

J

K

L

M

O

P

Q

R

S

T

W

X

Y

Z